CAMBRIDGE LIBRARY COLLECTION

Books of enduring scholarly value

Life Sciences

Until the nineteenth century, the various subjects now known as the life sciences were regarded either as arcane studies which had little impact on ordinary daily life, or as a genteel hobby for the leisured classes. The increasing academic rigour and systematisation brought to the study of botany, zoology and other disciplines, and their adoption in university curricula, are reflected in the books reissued in this series.

A Biographical Sketch of the Late William George Maton. M.D.

English physician William George Maton (1774–1835) was a polymath who had a special interest in botany: a shell and a parrot were among species named in his honour. His writings on natural history included a catalogue of the plant and animal life around Salisbury, Wiltshire, which was published posthumously in 1843 and is reissued as the second part of this composite work. The first part contains a sketch of Maton's life and work by fellow physician and writer John Ayrton Paris (c. 1785–1856), first presented to the Royal College of Physicians, and subsequently published in 1838. Paris discusses Maton's early life, his contributions to the growing field of botany, his other scientific and antiquarian interests, and his distinguished medical career, during which he was appointed physician-extraordinary to Queen Charlotte, wife of George III, and later physician-in-ordinary to the duchess of Kent and the young Princess (later Queen) Victoria.

Cambridge University Press has long been a pioneer in the reissuing of out-of-print titles from its own backlist, producing digital reprints of books that are still sought after by scholars and students but could not be reprinted economically using traditional technology. The Cambridge Library Collection extends this activity to a wider range of books which are still of importance to researchers and professionals, either for the source material they contain, or as landmarks in the history of their academic discipline.

Drawing from the world-renowned collections in the Cambridge University Library, and guided by the advice of experts in each subject area, Cambridge University Press is using state-of-the-art scanning machines in its own Printing House to capture the content of each book selected for inclusion. The files are processed to give a consistently clear, crisp image, and the books finished to the high quality standard for which the Press is recognised around the world. The latest print-on-demand technology ensures that the books will remain available indefinitely, and that orders for single or multiple copies can quickly be supplied.

The Cambridge Library Collection will bring back to life books of enduring scholarly value (including out-of-copyright works originally issued by other publishers) across a wide range of disciplines in the humanities and social sciences and in science and technology.

A Biographical Sketch of the Late William George Maton. M.D.

Read at an Evening Meeting of the College of Physicians

John Ayrton Paris

CAMBRIDGE UNIVERSITY PRESS

Cambridge, New York, Melbourne, Madrid, Cape Town,
Singapore, São Paolo, Delhi, Tokyo, Mexico City

Published in the United States of America by Cambridge University Press, New York

www.cambridge.org
Information on this title: www.cambridge.org/9781108038157

This edition first published 1838
This digitally printed version 2011

ISBN 978-1-108-03815-7 Paperback

A BIOGRAPHICAL SKETCH

OF THE LATE

WILLIAM GEORGE MATON, M.D.

FELLOW OF THE ROYAL AND ANTIQUARIAN SOCIETIES,
VICE-PRESIDENT OF THE LINNEAN SOCIETY,
AND FELLOW OF THE ROYAL COLLEGE OF PHYSICIANS.

READ AT AN EVENING MEETING OF THE
COLLEGE OF PHYSICIANS.

BY

JOHN AYRTON PARIS, M.D. F.R.S.
FELLOW OF THE COLLEGE.

LONDON:
RICHARD AND JOHN EDWARD TAYLOR,
RED LION COURT, FLEET STREET.
1838.

PRINTED FOR DISTRIBUTION BY THE EXECUTORS, WITH THE PERMISSION OF THE AUTHOR.

A

BIOGRAPHICAL SKETCH

OF THE LATE

WILLIAM GEORGE MATON, M.D.

WILLIAM GEORGE MATON was born at Salisbury on the 31st of January, 1774. His father, George, was a wine-merchant of considerable repute in that city, and having served the highest municipal office, was usually addressed as " Mr. Chamberlain Maton :" although far from affluent, he bestowed a liberal education upon his children. The subject of the present memoir was the eldest of *four*, all of whom he survived. His elementary education was obtained at the Free Grammar School of his native place, and it also appears that he was early initiated in the rudiments of Natural History, since a passion for scientific pursuits, even while a schoolboy, soon displayed itself, and is said to have considerably interfered with the progress of his more legitimate studies. This predilection was much strengthened by several concurring circumstances. He had not attained his tenth year, when it was his good fortune to have attracted the notice and fostering regard of the Reverend Thomas Rackett of Spetisbury, a gentleman too well known in the circles of scientific and literary distinction to require from me any further notice than an expression of acknowledgement and thanks for much of the valuable information which is contained in the present memoir. Some time afterwards he formed an acquaintance with that distinguished chemist and philosopher, Mr. Charles Hatchett, which, in the progress of time, ripened into a friendship that terminated only in the grave. The former of these gentlemen introduced his young friend to Dr. Richard Pulteney of Blandford, a physician of considerable eminence in the West of England, and the learned author of various works on Natural History, and, more especially, of one entitled "*A General View of the Writings of Linnæus.*"

In July, 1790, he was admitted at Queen's College, Oxford, as a commoner, and shortly afterwards he added to his scientific acquaintance Mr. Aylmer Bourke Lambert, a prominent name in the annals of botany, and Dr. John Sibthorp, Professor of Botany, and author of the "*Flora*

Oxoniensis" and "*Flora Græca*." It is fair to conclude that the constant and intimate intercourse which he enjoyed with these distinguished persons had a material influence in moulding his mind, and in inspiring it with a generous and noble emulation; indeed, we learn that in the preparation of both the works of Dr. Sibthorp, MATON was a zealous assistant. During the composition of the "*Flora Oxoniensis*," he attended the Professor in his herbarizing excursions; and although, from motives of prudence, he resisted tempting solicitations to accompany him to Greece, he nevertheless corresponded with him on subjects connected with the scientific objects of his travels. With the other eminent persons above-mentioned we shall hereafter find that he not only continued through life to cherish a warm friendship, but that his name has become honourably associated with theirs in the history of science; and, let me here observe, that although botany, to borrow the metaphor of Lord Bacon, was to him as "a first-born child, yet he did not make it his heir, to the exclusion of every other;" indeed, I have reasons for believing that, had he not been influenced by professional considerations, the study of conchology would have enjoyed a preference.

During the Oxford vacations he generally

visited London, for the purpose of mixing with the society of scientific men, and of attending the meetings of philosophical bodies.

It was on the 18th of March, 1794, when only in his twenty-first year, that he was elected a Fellow of the Linnean Society, which had the effect of introducing him to a wider circle of naturalists, all bound together by a common attachment to that distinguished botanist and most amiable man Sir James Edward Smith. We have only to search the Transactions of that learned body, to discover ample evidence of the zeal and industry with which MATON laboured to advance the progress of his favourite science. In the third volume, we shall find a paper, read before the society in 1794, describing a new species of *Tellina*, not noticed by Linnæus; the shell was found on chalky parts of the bed of the river Avon, and in rivulets communicating with it near Salisbury, and hence he gave it the name of *Tellina rivalis.* In the fifth volume of the same work is a paper entitled "*Observations on the Orcheston Long Grass.*" The object of this communication was to prove that the long grass of this celebrated meadow of Orcheston Saint Mary is not only not a species peculiar to the spot, as botanists had asserted, but that it is composed of most of the species which grow in other meadows, the luxuriance being favoured by the operation of several local causes. In the seventh volume, we are presented with a joint paper by Maton and his early friend Mr. Rackett, entitled "*An Historical Account of Testaceological Writers*," which occupies no less than 125 quarto pages of the Transactions, including a history of the most important labourers in this province of natural history, from Aristotle to the most modern writer. On this occasion, Maton pays a just tribute to the admirable researches of his friend Mr. Hatchett. He distinguishes him as being the only author, with whose writings we are acquainted, who has scientifically investigated the chemical character of shells, a comparison of which with those derived from external structure cannot but be highly curious and interesting to the philosophical naturalist. To the disciple of Linnæus, it is peculiarly satisfactory to perceive that so many of Mr. Hatchett's experiments tend to establish the propriety of distinctions adopted by that illustrious naturalist. "We need only," says Maton, "refer to the instance of the *Echinus*, the chemical characteristic of which genus proves, in opposition to Klein, the correctness of Linnæus in placing it among the *crustaceous* instead of the testaceous tribe; for the presence of the *phosphate of lime*, as detected by Mr. Hatchett in the covering of the *Echinus*, at once distinguishes the latter from *testaceous* bodies, which alone consist of *carbonate of lime*, mixed with gelatinous matter." It is with great satisfaction that I allude to this passage in the memoir, not only as furnishing a most beautiful example of the intimate connection that subsists between all the various branches of natural knowledge, by which one science, however distinct and apparently unconnected, can be thus made to shed a reflected lustre on others; but because I feel that, in our rapid progress along the stream of discovery, we are too apt to overlook the services of those who first set our bark afloat; I therefore rejoice in this opportunity of recalling to the recollection of my scientific brethren the early services of Mr. Hatchett in a new and unexplored region of philosophy. This paper is succeeded by another, equally elaborate and important, by the same author, being "*A Descriptive Catalogue of the British Testacea*," and containing an account of some remarkable shells found in cavities of a calcareous stone, called by the stonemasons "*Plymouth Rag*." In the tenth volume of the Transactions, MATON has described seven new species of *Testacea* from Rio de Plata, which had been placed in his hands by Sir Joseph Banks. In the same volume we shall find a learned critical and botanical note, appended by him to a paper on the *Cardamom* by Mr. White, surgeon, of Bombay, in which he designated the genus of which it consists *Elettaria,* a nomenclature to which Sir James Edward Smith objected, and proposed that of *Matonia*; a suggestion which was adopted by Mr. Roscoe in his description of the "*Scitamineæ*:" this, however, has been since superseded, on the authority of Roxburgh, by that of "*Alpinia Cardamomum.*"

The Linnean Society, as it would appear, duly appreciated the value of his services; his name was in the list submitted to the Crown for insertion in the charter granted to that body in 1802. He was annually re-elected into the Council, the President as repeatedly nominating him to the office of Vice-President. It ought also to be stated that the Linnean club acknowledges Dr. MATON as its founder and most constant attendant.

From the limited number, as well as the retired habits, of the true votaries of this elegant science, the researches of MATON can never become the objects of popular admiration, nor will they attract the regard of those who are engaged in the busy pursuits of life; but his kindred labourers have inscribed his name in a temple where his spirit best loved to dwell, not amidst the tumultuous intercourse of men, but in the deep and sequestered recesses of nature. Here will his memory be cherished in unostentatious simplicity, when the gorgeous monument and the storied urn shall have crumbled into dust; for although the memorial of the naturalist may be limited as to the extent of its publicity, it at least receives a compensation in the durability of its existence. Augustus Cæsar raised a statue to his physician Musa; Juba, King of Libya, evinced his gratitude for similar services by naming a plant after Euphorbus (*Euphorbia*), "*Ubi jam Musæ statua? Periit—evanuit! Euphorbii autem nomen perdurat—perennat.*"

By various members of the Linnean Society has the name of MATON been identified with objects of natural history; for instance, it has been given to a species of *Tellina* in Wood's *Index Testaceologicus*; to a genus of plants by Sir James Edward Smith, and afterwards by his friend Robert Brown in the *Plantæ Asiaticæ Rariores* of Dr. Wallich; and to a new species of *Psittacus* by Mr. Vigors and Dr. Horsfield, in their description of the Australian birds in the collection of the Linnean Society, and which is thus announced: "*In honorem* GULIELMI MATON,

Medicinæ Doctoris, Societatis Linneanæ PRO-PRÆSIDIS, *in Zoologia eximii judicis, hæc species pulchra generis pulcherrimi nomine distinguatur.*"

It would appear that, in the earlier part of MATON's life, his attention had been attracted to vegetable chemistry, as the means of occasionally confirming the distinctions of the naturalist, upon the same principle that the researches of Hatchett had become subservient to the views of the conchologist; and it is not a little remarkable that, in the prosecution of such an inquiry, he should have actually discovered the alkaloid principle of bark. By referring to the fifth volume of the London Medical and Physical Journal (p. 33), we shall find an account of the experiments, and of the precipitate occasioned by an infusion of nut-galls in the decoctions of bark; a fact which occasioned considerable speculation at the time, and which ultimately led to the detection of the active principle of that heroic remedy.

It would have been extraordinary had not the active mind of MATON been early attracted to those venerable remains of antiquity which give so romantic an interest to the vicinity of his native city; hence, having acquired a taste for antiquarian researches, he united their pursuit with the investigations of natural history. Nor are these studies, however dissimilar they may appear on a superficial view of their tendency, so far incompatible with each other as to require for their successful investigation minds of a different order and construction. On the contrary, the mouldering monuments of past ages, and the fresh and fragrant flower expanding its petals to the sun, although furnishing images of beautiful and vivid contrast, are connected, in the imagination of the philosopher, by trains of thought leading to the most interesting and profound reflections. We accordingly find MATON employing much of his leisure from severer studies in attempting to elucidate the history of Salisbury, by rescuing from oblivion the scattered records of the taste and feeling of former times. In the Gentleman's Magazine for 1792, he has given an account of a conventual seal found at Salisbury. He also contributed largely to the "*Salisbury Guide*," and to "*Hutchins's History of Dorset.*" He happens to have been the first person to whose lot it had fallen to record any facts connected with the history of the mysterious monument of Stonehenge; having described, in the thirteenth volume of the Archæologia, the circumstances attending the dilapidation which took place in 1794, and ascertained the weight and composition of one of its largest stones; by which he was enabled to settle a point that had occasioned considerable speculation and controversy. In the same year in which this paper was published, the Society of Antiquaries received him into their body at the age of 23, and repeatedly elected him into their Council. He also became early a Fellow of the Royal Society, and his name appears in the certificate as one of those who proposed Sir Humphry Davy for that honour.

I shall now record a circumstance of some interest that occurred in the year 1793. Dr. Maton, while resident in Oxford, established a small society for the advantage of scientific and literary discussion. The heads of houses, however, denounced this learned confederacy as a dangerous encroachment upon the statutes, and, although all political questions were expressly excluded by their laws, they were not allowed to hold their meetings in any public room. The society was subsequently remodelled, and transferred to the more congenial soil of the metropolis, under the name of the "Academical Society;" its founder, Dr. Maton, having for several successive years been elected as President. The majority, however, of the new members being students in law, it became subservient to the cultivation of oratorical talent, and the original members ultimately seceded; among whom were Sir John Stoddart, President of the Supreme Tribunal of Justice at

Malta; Lord Moncrieff, one of the Lords of Session in Scotland; Dr. Copleston, the learned Bishop of Llandaff; and Dr. Dibdin, the bibliographer. To these acceded, when the society was re-established in London, the present Right Hon. Lord Glenelg and Robert Grant; Sir Launcelot Shadwell, the Vice-Chancellor of England; Sir Benjamin Brodie, Bart.; Sir Henry Ellis, the learned Secretary of the Society of Antiquaries; Mr. Justice Williams; Sir John Campbell, the present Attorney-General; Mr. Serjeant Storkes; the late Mr. Bowdler; Lord Gifford; Dr. Reeve, and Dr. Bateman. Amongst the frequent visitors who took part in the discussions, were Lord Brougham and the present Lord Advocate of Scotland.

In the summer of 1794, Mr. Hatchett and Mr. Rackett, having arranged to make a tour into Cornwall, invited their young friend MATON to accompany them. An opportunity of thus gratifying his spirit of inquiry, and, more particularly, of pursuing the studies of mineralogy and geology, held out a prospect of intellectual pleasure that was irresistible. Mr. Hatchett, as we so well know, was the first chemist and mineralogist of the day; Mr. Rackett was not only a distinguished naturalist, but a zealous and enlightened antiquary. With such able companions, it would indeed have been extraordinary had not MATON realized all the advantages his most sanguine hopes could have anticipated; and let it be remembered that, at this period, neither the mineralogical productions, nor the geological structure of that important district of our empire had received any remarkable attention. The spirit of Davy had not summoned from the dark recesses of its mines and caverns those genii which have latterly displayed the extent and variety of their subterranean treasures. Upon this subject I feel that I am competent to offer an opinion; and I assert, without the fear of contradiction, that MATON was the first scientific labourer in that extensive field of mineral riches. During his tour he entered in his journal every phænomenon connected with the mineralogy, geology, antiquities, and natural history of the districts he visited, while his friend Mr. Rackett was occupied in representing the more striking beauties of the scenery by a series of masterly sketches. On the termination of this tour, which included considerable portions of the counties of Dorset, Devon, and Somerset, besides Cornwall, which may be said to have been its more immediate object of attraction, Dr. MATON found that he had collected such a mass of interesting and novel information as to sanction the publication of a distinct work. The memoranda, however, were first submitted to Sir Joseph Banks, who most strenuously supported him in his laudable design, while his friend Mr. Rackett placed at his disposal such drawings as might be considered necessary for its illustration. To enter into an analysis of this elegant production of the early years of my lamented friend, would carry me far beyond the bounds to which I am necessarily limited. The antiquary, the geologist, the mineralogist, the conchologist, and the historian may each derive satisfaction and instruction from its pages. I must not, however, dismiss the subject without remarking, that the work is accompanied by a plan, so ingeniously engraved, as to represent the various rocks and sub-soils of which the country consists; and, let me add, that this was the first attempt in England to construct A GEOLOGICAL MAP.

Dr. MATON was originally intended for the clerical profession; and, in truth, from the natural timidity and reserve of his disposition—from his romantic delight in the luxuries of retirement, and in the charms of a country life, it is a question whether such was not the path, had he consulted his individual happiness, which he ought to have pursued; he seems, however, somewhat hastily to have abandoned his original intention. All his favourite studies connected themselves with the science of medicine, and the influence of his friend Dr. Pulteney, to whom he had now leisure to make more extended visits, without doubt, fostered his growing attachment to it. Having, therefore, obtained the concurrence of his father, he commenced his medical studies in the spring of 1797, by entering himself as a pupil at the Westminster Hospital, and to various lecturers. Dr. Baillie was at that time the most distinguished anatomical teacher in the western part of the metropolis, with whom was associated the ingenious Mr. Cruickshank. On the admirable lectures of these eminent persons he was a constant and diligent attendant, and with the former an intimacy arose which continued to the end of the life of that most exemplary man and distinguished physician. It was chiefly from the encouragement and advice of Baillie that he acquired the firmness and perseverance which enabled him to encounter the difficulties, and to sustain the exertions, during the early years of his professional career. To his sound counsels, and to his pure example, he was deeply indebted for the due regulation of his conduct as a member of an enlightened profession and as a man of the world. Having thus accomplished himself in all the collateral branches of medical science, and been initiated in the practice of physic by a sedulous attention to the phænomena of disease in the hospitals of the metropolis, he commenced the actual practice of his profession in the year 1800, having been previously admitted to the degree of bachelor in medicine. In 1801 he proceeded to the doctorate, and was in due course elected into the fellowship of the college; and confidently may I appeal to his brethren, whether, in all its social relations, its high professional duties, and in its most sacred obligations, he did not acquit himself with zeal, talent, firmness, and integrity. He discharged, *seriatim*, all the offices which, as Fellow of the College, he was called upon to perform, as Gulstonian Lecturer, Harveian Orator, Censor, and Treasurer, and lastly as an Elect. Nor must we pass over the essential services he rendered the College and the profession by his assistance in arranging the nomenclature of the botanical and zoological articles of the Materia Medica in three successive editions of the Pharmacopœia. To the Transactions of the College, he was also a contributor. In the fourth volume he has a communication on the obscure subject of *Superfœtation.* In the fifth are two papers; the one entitled "*Some Account of a Rash liable to be mistaken for Scarlatina*;" the other relates a case of Chorea in an aged person, cured by musk.

Very shortly after he had attained the Doctor's degree, he was unanimously elected physician to the Westminster Hospital, on the resignation of Dr., now Sir Alexander, Crichton. This to so young a practitioner was an appointment of great importance; but, early as his professional career commenced, we may conclude that before the period alluded to, his acquirements were duly appreciated by his contemporaries, for after reading and defending a paper upon *Chorea*, we find him seated as one of the Presidents of the "Lyceum Medicum," a society established by John Hunter, and which was at that time the popular arena for medical discussion. The first few years, however, of his practice were, as may well be supposed, all but unproductive; and his pecuniary resources being inadequate to sustain a long state of expectancy, he was advised to adopt a system, not unusual with the young metropolitan physician, that of residing at some popular watering-place during the season. Weymouth was suggested to him as a place well calculated to respond to his wishes; and as this important period of his life is marked by circumstances of considerable interest, I have endeavoured to collect the most authentic particulars relating to it. From Robert Benson, Esq., of Lincoln's Inn, I have derived the following statement: "My father," says he, "who knew Maton from his infancy, and always expressed a high degree of interest in his professional success, accompanied him in his first journey to Weymouth. Having at that time little or no practice, Maton had ample leisure to pursue his botanical researches, and such was the zeal and diligence which characterized his pursuit, that his rambles about Weymouth attracted very general notice. Their Majesties, it will be remembered, were passing the season at Gloucester Lodge, and one of the Princesses amused herself with botany. It so happened that a plant was brought to the Royal student, not uncommon in the neighbourhood, but which was unknown to Her Royal Highness; it was the *Arundo Epigejos*. Dr. Maton was mentioned to the Queen as a person likely to solve the difficulty; and as my father," continues Mr. Benson, " was accidentally strolling with Maton along the Esplanade, an equery of Her Majesty came from the Lodge, and addressing himself to the former, informed him that Her Majesty desired to see him. My father was, as may be readily imagined, not a little astonished at this announcement; and the equery, perceiving, no doubt, some hesitation in his manner, inquired whether he was not addressing Dr. Maton. The mistake was at once explained, and the Doctor accompanied his conductor to the Presence. Such was the origin of his introduction to the Royal

Family, to which he was unquestionably much indebted for his early advancement to profitable practice. It gave him a name and character at Weymouth highly advantageous to his profes sional views; and the manner in which George the Third subsequently mentioned his talents and acquirements, at once secured for the young physician the confidence of all the courtly invalids who required the aid of superior advice; and it led, in 1816, to his appointment as Physician Extraordinary to the person of Her Majesty, Queen Charlotte. The Duke of Kent having been attacked with a serious illness in Devonshire in 1820, it was thought advisable to call into consultation a London physician, and Dr. MATON was selected upon this responsible occasion; and, although his efforts, in conjunction with those of the domestic physician, failed in saving the life of the illustrious Duke, still his zeal and attention were duly appreciated, and served to cement his connection with the different branches of the Royal Family; and Her Royal Highness the Duchess of Kent, without any solicitation on his part, appointed him Physician in Ordinary to herself, and to her Royal infant the Princess Victoria; and to the hour of his death he was regarded, not only as the confidential physician, but as the private friend, and the arbiter of all scientific questions of interest which presented themselves to the notice of that Royal circle."

I must now carry you back to the year 1805, an important æra in the scientific life of MATON. His early acquaintance with Dr. Pulteney, and the more mature friendship in which they were united, have been already mentioned. On the death of this distinguished naturalist, MATON, to whom he had bequeathed all his botanical manuscripts, conceived the plan of editing his "*General View of the Writings of Linnæus*," and of prefixing a biographical memoir of its departed author, "whose well-merited eminence," says he, "both as an author and as a physician, seemed likely to render it acceptable to the public; while," he adds, that "he felt an additional motive to pay this tribute to his memory, in the grateful remembrance of a friendship which influenced his pursuits at a very youthful period, and to which he was indebted for many of the most instructive and agreeable hours of his life." In the execution of this work, accident put him in possession of some documents, which not only contributed to enhance its value, but to bestow upon it the claims of originality. These consisted of a diary of Linnæus, written by himself, and which was now for the first time translated into English from the Swedish manuscript.

I have now to record another work, or rather an appendix to that which Dr. MATON designates as one of the most superb offerings at the altar of Flora ever made by a private individual, Mr. Lambert's "*Description of the Genus Pinus*;" to which MATON contributed "An Account of the Medicinal and other Uses of various Substances prepared from Trees of that Genus." These products, both native and artificial, are much employed in medicine and the arts; and the terms commonly attached to them are, in general, extremely vague, ambiguous, and inexpressive. In this appendix, it is the author's object to dissipate the confusion, by substituting appropriate appellations for those which are either ambiguous or likely to lead to error, and by immediately arranging under every head such synonyms as may be adduced without undue latitude of conjecture.

In the year 1809 his private practice had become so extensive, that he found it necessary to retire from the office of physician to the Westminster Hospital, and, through his kind recommendation and patronage, I had the good fortune to succeed him in that important situation. During the latter fifteen years of his life, the reputation he had acquired brought with it such a pressure of professional labour, that it became absolutely necessary for the preservation of his health that he should devote several weeks of the autumn to relaxation and a total abstraction from business. These brief, but bright intervals were employed in visiting the interesting districts of his own country, as well as various parts of the continent of Europe. France, Switzerland, Germany, Venice, Rome, Naples, and the Pyrenees successively afforded him objects of novel inquiry and of philosophical contemplation; and were I not limited by time and space, I might select from his various letters such passages as would not only indicate the elevated delight he derived from these tours, but exhibit him as an animated and most engaging correspondent.

The person of Dr. MATON is too well known to this audience to require any description; I shall therefore merely relate an anecdote, communicated to me by Mr. Rackett. He states that the secretary to the late Pretender Charles Edward, Mr. Lumsden, once observed to Dr. MATON at a dinner party, that he could not keep his eyes from him, so strongly did he resemble his master when in the prime of life.

In private life, no man in his intercourse with society was more agreeable in his manners, or more sincere and steady in his friendships—no one more charitable and benevolent in his disposition: his notion of honour was refined to the extent of chivalry; his affection for his relatives and kindred unbounded, and his generosity towards them was only exceeded by the high sense of integrity which occasionally led him to exercise it. In short, I know not whether it is my admiration and esteem of his character, the remembrance of an uninterrupted friendship of thirty years, and the grief I experience for his loss,—I know not whether these circumstances have bribed my heart and blinded my judgement; but I cannot contemplate the assemblage of virtues in my departed friend, without regarding him as having approached almost as nearly to perfection as the frailties of our nature can allow. Nor should I do his memory full justice were I to pass unnoticed a noble act of beneficence, alike uncommon in the extent of the sacrifices it demanded and in the circumstances which induced it. The motive was unconnected with any selfish gratification, anticipated recompense, or prospective advantage; it was the sole offspring of a love of justice, a stern determination to sustain what is right, at any sacrifice of personal comfort or even of worldly prudence. On the death of his father, in the year 1816, the latter years of whose life had been embittered by protracted bodily suffering, which had the effect of throwing all his accounts into confusion and arrear, a large unexplained balance was found due from his estate, after applying all his available assets. Thus MATON, instead of inheriting considerable property, as he had every right to expect, and which, on the threshold as he then was of his profession, would have been most important to him, unexpectedly found himself called upon to administer an insolvent estate, and to provide for his collateral relatives who had depended upon his father for support, or who had, in the declining age of both his parents, afforded them the necessary attentions and comfort. His resolution was immediately formed; he prevailed upon every creditor to accept his debt by instalments; and, in order that he might faithfully redeem the pledge he had given to them, he annually set apart such a portion of his income as he could spare, after defraying the expenses which were essential to his professional station and appearance. At length he achieved his noble object; he liquidated the debts of his father, and he provided for those who were dependent upon him; but it was through long toil, anxiety, and a secret depression that weighed on his sensitive mind that he accomplished it. I am informed that a sum exceeding 20,000*l.* was for these purposes expended during his life; and thus were the means by which large fortunes are generally created—regular accumulations upon early savings—devoted to purposes far more sacred, and more gratifying to a mind, actuated at all times by a moral feeling, devoid of every selfish and sordid alloy.

It would have been "passing strange" had not the citizens of his native place justly and gratefully appreciated so noble an act of honourable disinterestedness; and they accordingly recorded their sense of his character by a civic memorial. The mayor and corporation presented him with the freedom of the city, in a splendid gold box, bearing the following inscription:

THE MAYOR
AND COMMONALTY OF NEW SARUM,
TO
WILLIAM GEORGE MATON, M.D.,
F.R.S., F.L.S., F.A.S.,
WITH THE FREEDOM OF HIS NATIVE CITY,
TO MARK THEIR ESTEEM FOR HIS TALENTS AND CHARACTER.
1827.

Dr. Maton had, through life, anticipated with much pleasure the period at which he might rest from his professional labours, and retire into the depths of the country to indulge in his favourite pursuits. About a year before his death he had so far advanced his projected plan, as to become the purchaser of Redlynch House, near Downton, in Wiltshire, where he spent some time during the autumn. Upon this occasion a circumstance occurred, as if his presiding genius had contrived to pamper his ruling passion, and heighten the joy of his anticipations. He discovered growing near his domain the *Asarum Europæum*, one of our rarest British plants. On his return to town, he frequently spoke to me of the pleasure and satisfaction he had received from his autumnal retreat, and expressed his determination to indulge in that retirement every succeeding season. It also appears that he had arranged some plans which were to afford him some agreeable occupation during these intervals of relaxation from the severer duties of his profession. From a letter which I have received from Mr. Benson, it would appear that his taste for antiquarian researches, which so greatly distinguished the period of his youth, had revived during his declining days, for he informs me that he had undertaken to collect materials for Sir Richard Hoare respecting the history of Salisbury and its vicinity; and that Dr. Maton had assisted him by the production of notes made in early life, from which he had obtained a complete catalogue of the plants indigenous to a circular area of twelve miles in extent round that city. By the same authority I am informed that in 1831 Dr. Maton suggested the formation of a Society to consist of Wiltshire topographers, each of whom was to take up a different station in the county during the autumn, with a view to examine, personally, the various objects of interest in each district. The plan, however, was never carried into execution. Mr. Benson was consigned to the bed of sickness; MATON terminated his career in the grave.

The precarious tenure of our existence, and the vanity of human wishes, suggest a sentiment so trite, and so destitute of grace and novelty, as to expose the writer who would venture to moralize upon its tendency to the charge of a commonplace declaimer. There is, however, something so striking and instructive in all the circumstances by which it is exemplified in the instance before us, that I cannot pass it over without a comment. To how many of us assembled here may it not furnish a salutary lesson!—to all, if properly considered, it will operate as the means of extinguishing the thirst of avarice, of assuaging the fervour of ambition, of moderating the too eager aspirations of hope, and of soothing the bitterness of disappointment. Six months had scarcely elapsed, before MATON, who had been thus exhilarated by the prospects of happiness, and engaged in maturing plans for years of intellectual enjoyment, is abruptly removed from the bright scenes of his anticipation by an unexpected and painful death.

It is true that his malady had been stealing upon him for several years, but so gradual had been its invasion as to have been almost imperceptible to himself, so that the rapidity with which it progressed during the last few weeks, gave to its termination all the afflicting characters of an unexpected and sudden death. He died at his house in Spring Gardens on the 30th of March, 1835.

Notwithstanding the length to which this memoir has been extended, I trust it will not be necessary for me to offer any apology for the time I have occupied or the attention I have solicited. Let it be remembered, that the very object of these evening meetings of the College is to draw together the different members of our profession, not only for instruction, but with a view to encourage a sentiment of mutual good-will and kindly feeling,—to soften existing asperities by quiet collision and friendly intercourse,—and, in the language of Sterne, " to teach the milk of human kindness to flow all cheerily in gentle and uninterrupted channels," so that we may, one and all, cordially unite for the common purpose of raising the standard of our moral and social qualities. In pursuance of such an end, let me ask, whether a subject can be found better calculated to attain our object than a memoir of the accomplished physician and excellent man whom I have attempted to delineate? The very contemplation of such a character soothes every turbulent feeling, and harmonizes within us those elements which might otherwise be discordant, and, by putting us in better humour with ourselves and with our profession, it must incline us to entertain a kinder regard for the intentions of each other. Nor is the worldly success and temporal reward which crowned the career of MATON without its moral. It has too generally been asserted, that the predilection which the public have evinced for particular physicians can rarely be traced to any acknowledged principle connected with the possession of substantial merit, or with the honourable exercise of superior talent. It must, however, be acknowledged, that the instance of Dr. MATON, amongst many others recorded in the annals of our College, offers a striking refutation of so humiliating a proposition. No one more anxiously desired to divest his profession of every selfish and sordid consideration: he had early enlisted himself under the banners of truth, and sooner would he have forfeited every chance of promotion than have rested his hopes of success on an unholy alliance with the spirit of delusion. It is true that he treated the prejudices of his patients with indulgence and regard, but his professional advancement was never marked by a mean submission or a servile attention to their wishes, nor by an abject homage to their rank or opulence. He won their confidence by a distinguishing sagacity and a prompt judgement, manifested in a manner at once decisive, but unaffectedly courteous and engaging. He maintained this advantage by the success of his treatment, and by the warm and active diligence with which he directed it.

THE END.

PRINTED BY RICHARD AND JOHN E. TAYLOR,
RED LION COURT, FLEET STREET.

THE

NATURAL HISTORY

OF A PART OF

THE COUNTY OF WILTS,

COMPREHENDED WITHIN THE DISTANCE OF TEN MILES ROUND THE CITY OF SALISBURY.

BY THE LATE

GEORGE MATON, M.D.

F.R.S., V.P.L.S.

LONDON:

PRINTED BY J. B. NICHOLS AND SON,

25, PARLIAMENT STREET.

1843.

INTRODUCTION.

To these Notices it is proper to premise, that most of them were taken at an early period of the writer's life, and before the commencement of the present century, so that, as far as Botany is concerned, the progress and changes of cultivation, and the alteration of various influences which must have happened in the course of so many subsequent years, cannot but have destroyed the productiveness, as to particular species of plants, in some spots, and furnished *pabulum* for different species, and perhaps more remarkable ones, in others. Indeed, practical botanists have not unfrequent occasion to remark that, even where no appreciable alteration of soil or influence has occurred, the same race of plants is not always continued; and, on the other hand, from causes equally obscure, strangers will begin occupation of the spots deserted by the *aborigines.* A collector therefore will often be disappointed, and often surprised, on visiting *loci natales*, to which he may have been directed by the compiler of a botanical catalogue,—disappointed at not finding a single specimen of what may even have been described as growing in a particular place *most luxuriantly* (the season of the year the same, and the condition of the ground apparently unaltered), and surprised at the sight of some other species there, for which he was not prepared, and which he may therefore be apt to think his author had overlooked. In illustration of the preceding observations, the writer cannot help mentioning an unaccountable fact in regard to that elegant and unfrequent plant, *Neottia spiralis*, which in the month of August, in the year 1813, he found growing over a surface many yards square; and with its white blossoms so numerous as to assume, at a distance, the appearance of an immense sheet; yet, in no succeeding year has he been able (notwithstanding the season and the state of the ground were the same,) to discover, in or near that spot, a single specimen of it. Whatever ravages browzing animals or insects may commit on vegetable productions, it is difficult to conceive how an entire and large generation of the latter can be *wholly* eradicated by the former in a single year; and how great soever may be the variation of condition in the atmosphere in different seasons, it is most natural to suppose, that in a *series* of years the same combination of influences favourable to the germination of a particular plant, once an inhabitant of a particular soil, must come round again; but that plant, it would seem, may nevertheless become extinct there for ever. Something analogous to this is observable in regard to the animal kingdom. It is well known that large shoals of the herring may frequent, for several consecutive seasons, certain parts of the coast, and suddenly this visitation will cease for as many, or more seasons, not even a straggling individual being seen—yet no cause for this shall be cognizable. If it should be said, that the Herring does not find food always in the same place, this is only substituting one difficulty for another. As to the total disappearance of *some* animals, indeed, from a region once well stocked with them, it may be accounted for easily. The Bustard, for instance, was formerly an inhabitant of Salisbury Plain,* but the writer believes that one might as successfully look for an ostrich there in the present day. Now, it is almost superfluous to remark, that not only has cultivation closely gained upon the haunts of this bird, and thus curtailed him of his appropriate food, but human habitations have also disturbed his solitudes, so that such an abode can no longer suit his nature.

There are few inland districts perhaps (of the same extent) in which a botanist will find a greater variety of plants than in that which forms the subject of the present catalogue,—and which not only comprehends meadow, marshy, arable, wood, and waste land, but also has for its sub-soil many different substances, as chalk, limestone, clay, gravel, and sandstone. It includes lofty hills, spacious irrigated vallies, extensive corn-fields (both of upland and plain), and open downs. By far the greatest part, however, consists of chalky eminences; hence, many species are common to it with the contiguous counties of Hants and Dorset, though some would seem to inhabit Wilts exclusively. The writer's opportunities of observation have been by far most frequent in the immediate vicinity of Salis-

* A very observant and credible person of the name of Dew, whom I knew as a sportsman in my younger days, informed me in the year 1796, that he once saw as many as seven or eight of these birds together on the Downs, near Winterbourn Stoke; but I have not met with any one since who has actually *seen* the bustard in Wiltshire subsequently to that year.

bury (his native city), and of Downton; but as there have been parts of the year at which, in early life, he was generally resident at the University, and, in later years, confined by the duties of his profession to the metropolis, it is probable that many remarkable plants have (on account of their season of flowering) escaped his notice; *some*, indeed, which even have places in this catalogue, have not actually been seen by himself, but he was unwilling to exclude from it any subjects rare or curious, when he was fortunate enough to have sound authorities for the insertion of them. In the cases alluded to, the names of the authorities will be found attached to the several species, respectively. And here he is impelled to speak somewhat particularly of such fellow naturalists as have trodden the same ground as himself, with the same desire to ascertain the zoological and vegetable productions of his native county; of such at least as are hitherto known to him, either by their works, or by their personal communications with him. First and foremost among these must be placed the venerable name of John Aubrey, F.R.S., whose "Memoires of Naturall Remarques in the County of Wilts," may be found in a large MS. volume, in the Library of the Royal Society, and also, I believe, in the Ashmolean Museum at Oxford. They are dated "1685," and dedicated to Thomas, Earl of Pembroke. From the Preface it would appear (and there seems to be no reason for doubting) that Wiltshire has to boast of being the earliest county to find an historian of its productions.*

* "I am the first," says he, "that ever made an Essay of this kind for Wiltshire, and (for aught I know) in the nation, having begun it in anno 1656. In the year 1675, I came acquainted with

The history, it is true, includes a most heterogeneous mass of marvellous, crude, and (to the present generation) uninteresting matter; but interspersed in it are many curious, if not important, notices and remarks. Credulity was the common fault of that age; and it ought not to excite surprise that a work, even professing to be *scientific*, should, at such a period, be replete with opinions and narrations which must *now* be considered absolutely ridiculous and incredible. That illustrious naturalist, RAY, next deserves to be noticed as a contributor to the *Flora* of Wiltshire, through the medium of Gibson's edition of Camden's "Britannia;"* but his communications,

Dr. Robt. Plott, who had then his Naturall History of Oxfordshire upon the loome, which I seeing he did performe so excellently well, desired him to undertake Wiltshire, and I would give him all my papers, as he had also my papers of Surrey: and offered him my farther assistance. But he was then invited into Staffordshire, to illustrate that county, which having finished in December, 1684, I importuned him again to undertake this county, but he replied, that he was so taken up in (arranging?) of the Museum Ashmoleanum that he should meddle no more in that kind (unless it were for his native county of Kent), and therefore wished me to finish and publish what I had begun." Pp. 1 and 2.

* In this work (p. 114), particular mention is first made of the *Orcheston long-grass*, of which Ray says, "We are not yet satisfied what sort of grass this might be; and recommend the inquisition thereof to the industrious and skilful Herbarists of this Countrey." He calls it "Long trailing dog's grass," and speaks of its being found "by Mr. Tuckers, at Modington, some nine miles from Salisbury, with which they fat hogs, and which is four-and-twenty feet long." Very various and imperfect notices of this phenomenon in vegetation were given by succeeding writers. At length, however, the author of this catalogue had the satisfaction of clearing up all the difficulties relative to its true history, in a paper which he communicated to the

so far as they relate to this county, seem to have been made chiefly on the authority of other persons.

In the year 1799, Wiltshire was visited by two most zealous botanists, Mr. Dawson Turner and Mr. James Sowerby, who included it in their "Tour through the Western Counties of England," and described several of its vegetable productions in their catalogue of rare plants, printed in the 5th volume of the Linnean Transactions.

In 1817 a periodical work was begun by the late Henry Smith, M. D. of Salisbury, under the title of "*Flora Sarisburiensis*," which was intended to describe and illustrate, by coloured engravings, the English plants in general, but the author's attention was directed more particularly to those growing in the vicinity of Salisbury, and he included in his plan ample notices of their medicinal, economical, and other virtues; he was not, however, sufficiently encouraged in his laudable undertaking, and only four numbers of the *Flora* were published.

The above I believe to be the only authors whose attention was expressly given to the Natural History of Wiltshire, prior to the time of the collections made with that view by the writer of these pages, which were begun as early as the year 1792, though only very small portions have hitherto appeared in print, and these either in the Transactions of the Linnean Society, or in the works of contemporary indi-

Linnean Society, and which will be found in the 5th volume of the Transactions of that learned body. He found it to be not one particular grass, but *several* well-known and common species, remarkably elongated in consequence of local influences, which he ascertained by actual observation of the spot, in the year 1798.

viduals. But he has the pleasure of commemorating several other names, as being those of persons to whose scientific researches, in the district to which his own attention has been principally devoted, he is indebted for additions to his Catalogue, though the results had not been committed to the press by those persons themselves. Among these names the county of Wilts has to boast of one which not only belongs to the more ancient part of its native resident gentry, but is claimed by Science herself as that of one of the oldest, as well as most ardent, of her votaries now living,—that of AYLMER BOURKE LAMBERT, Esq. of Boyton House.*

The mention of this distinguished botanist will always be associated, in the mind of the writer of these pages, with some of the most delightful recollections of his life; for with him he made his earliest herbarizations in the environs of his native city, where several of the plants which he has recorded were first pointed out to him by the instructive companion of

* This venerable mansion ought to be recorded with particular respect in every work relating to the History of Wiltshire, its worthy possessor, whilst making it annually his abode in the summer, having been in the habit of hospitably receiving under its roof visitors of the highest celebrity in every department of useful knowledge; some of whom, whilst sojourners in it, have made curious and interesting discoveries connected with the Botany of the county, as also with its antiquities and works of art. Close to the precincts of Boyton House, one of the rarest of our British Plants, *Cyperús longus*, was found by William Peete, Esq. and to Mr. Lambert himself we owe the addition of *Coricus tuberosus* to that number; a species new to the British Flora. Mr. Lambert discovered the latter growing profusely in some marshy ground about two miles from his house.

his walks: with whom he has now enjoyed an uninterrupted friendship of more than forty years.

To not a few of the places of growth particularised in the present list of plants will be found subjoined the name of JOHN ROBERTS, A. L. S., who, though moving in a very humble rank of society,* was an indefatigable botanist, and discovered some interesting species to be natives of this district.

The latest collector of Wiltshire plants to whom this work is indebted for communications is ROBERT DICKSON, M.D., whose visits in the county having been only in the winter season, his researches were necessarily confined to the class *Cryptogamia.*

With regard to the arrangement of the several subjects of Zoology and Botany enumerated in the following Catalogue, it may be observed that, under each *order,* the *genera,* and, under each genus, the *species,* are distributed alphabetically, this plan appearing to be most convenient for reference, and the *systematic* place of each subject being ascertainable by consulting the authors who are quoted respectively. It was judged proper to adopt the Linnean nomenclature so far as it could be carried consistently with modern science and discoveries. In fact the Catalogue was made, *originally,* in that nomenclature *solely,* and the writer hopes to be pardoned if, as his heavy professional avocations do not allow him to keep pace with the daily improvements which are made in every

* This meritorious individual was a miller, and had been a member of a little club of persons occupying similar stations in life, who had associated for the express purpose of collecting and studying plants at Norwich. The writer's acquaintance with him began when the former was young both in years and in botanical knowledge.

branch of knowledge, some names now become almost obsolete, should not have undergone the corrections which it has been his wish to make as often as was requisite. He conceived that it would be useful to the reader to be referred, under each species, to some work in which a particular description of it might be found, and also to a figure of it (if likely to be accessible) for the purpose of identification. In the botanical part of the Catalogue, the "English Flora" of the late Sir James Edward Smith has been closely followed, as being the most approved and correct modern work, based on the system of Linneus: but, as its excellent author, unfortunately for science, did not live to render it complete by including the whole of the class *Cryptogamia*, the *Muscologia Britannica* of Drs. Hooker and Taylor, the *Synopsis Methodica Lichenum* of Professor Acharius, and other writers of the most established repute, have been followed wherever the Linnean generic divisions had been, by the common consent of Naturalists, superseded, or the species have proved defective or erroneous.

It will be evident, even on the most cursory view of the following pages, that it was not the object of the writer to compose anything like a complete *Fauna* or *Flora*. The notices were originally recorded solely for the private use of himself; they were made incidentally, and only when opportunities offered themselves; and, as authorship was not contemplated, they would never have assumed at all a regular form had there been any other person prepared to communicate to the historians of Wiltshire a systematic catalogue, however imperfect, of the natural productions of the county. The present contribution is offered merely as a *basis*, and that a scanty one, on which future observers may constitute an enlarged and useful *conspectus* of the Zoology and Phytology of the vicinity of Salisbury; and the writer hopes that those who succeed him will include *Mineralogy*, for which a very wide and curious field is open in this district, but into which (though he had been a zealous collector of fossils in his early days) he himself, not having been able to keep pace with modern science in that department, has not ventured to enter.

To conclude these prefatory remarks, it remains only to explain upon what principle so many species of plants, to be found almost everywhere, have been admitted into a catalogue from which so many, equally common, have been excluded. Of Cryptogamia species a large proportion, of very frequent occurrence, will be seen to be enumerated, though a large number of phænogamic, not even presenting themselves equally often, have been omitted. The explanation is, that the more common of the latter description of vegetable productions are too well known even to the vulgar to require being noticed in print; whereas many beautiful mosses, liverworts, &c. would, in consequence of their diminutiveness, never perhaps obtain the attention and investigation which they merit from a curious eye, unless they were included in a scientific catalogue.

Explanations of Abbreviations used in the following Catalogue.

Acharii Lich. Univ.—Lichenographia Universalis; ab Eric Achario, Eq. Aurat. M. D. &c. cum tabl. ær. col. Gotting. 1810, 4to.

Acharii Syn. Meth. Lich.—Synopsis Methodica Lichenum, ab Erik Achario, Eq. Aur. &c. Lundæ, 1824, 8vo. pp. 392.

Fries, Syst. Mycol.—Systema Mycologicum, sistens Fungorum Ordines, Genera, et Species huc usque cognitas, quas ad normam Methodi Naturalis determinavit, disposuit, atque descripsit Elias Fries. Vol. i. 8vo. Lundæ, 1821. Vol. ii. Sect. 1. Lundæ, 1822. Sect. 2, 1823, Gryphiswaldiæ.

Greville, Alg. Brit.—Algæ Britannicæ, or Descriptions of the Marine and other Inarticulated Plants of the British Islands belonging to the order of Algæ. By Robert Kaye Greville. 1830, 8vo. pp. 218, pl. col. 19.

Hooker and Taylor, Musc. Brit.—Muscologia Britannica. Ed. 2nd, 1827, 8vo. pp. 270, tabl. 30. With a Supplement of the *Hepaticæ*, tabl. 6.

Linn. Syst. Nat. à Gm.—Caroli à Linné Systema Naturæ, cura Joh. Frid. Gmelin, Phil. et Med. Doct. Ed. 13, Lipsiæ. 9 vol. 8vo, 1788—1793.

Linn. Tr.—Transactions of the Linnean Society of London, 4to.

Penn. Brit. Zool.—British Zoology, by Thomas Pennant. Ed. 4, 1776, 4 vols. 8vo.

Smith, Engl. Fl.—The English Flora, by Sir James Edward Smith, &c. &c. 4 vols. 8vo. 1824—1828.

Sowerby, E. B.—English Botany, or Coloured Figures of British Plants, with their Essential Characters, Synonyms, and Places of Growth, by James Sowerby. 36 vols. 8vo. 1790—1814.

Sowerby, E. B. Suppl.—Supplement to the above, by William Jackson Hooker, LL.D. and the fig. by James de Carle Sowerby. vol. i. 8vo. 1831.

Sowerby, Fungi.—Coloured Figures of English Fungi, by James Sowerby. 3 vols. fol. 1797—1803.

Wood, Ind. Test.—Index Testaceologicus, or a Catalogue of Shells, British and Foreign, &c. by William Wood. London, 1825, 8vo. with coloured engravings of every species.

VEGETABILIA.

"It were to be wished that we had a Survey or Inventory of the Plants of every county in England and Wales, as there is of Cambridgeshire by Mr. John Ray, that we might know our own stores, and whither to repair for them for medicinal uses."—Aubrey's MS. on Wilts, p. 116.

ACER—MAPLE.

A. campestre. Smith's English Botany, vol. ii. p. 231. Sowerby's English Botany, vol. v. pl. 304.

Hedges, not unfrequent. Very abundant in some closes between Downton and Redlynch.

ACHILLEA—YARROW.

A. Ptarmica. Smith, E. B. vol. iii. p. 461. Sowerby, E. B. vol. xi. pl. 757.

Meadows, near Laverstock and Stratford.

ADONIS—PHEASANT'S EYE.

A. *autumnalis.* Smith, E. B. vol. iii. p. 43. Sowerby, E. B. vol. v. pl. 308.

Corn-fields, near Pitton.

ADOXA—MOSCHATEL.

A. Moschatellina. Smith, E. B. vol. ii. p. 242. Sowerby, E. B. vol. vii. pl. 453.

Hedge-rows about Alderbury Hill, and on the hills above Laverstock.

AGARICUS—AGARIC.

A. campestris. (Mushroom.) Fries, Mycol. vol. i. p. 281. Sowerby, Fungi, pl. 305.

On downs and in pastures, common.

AGRIMONIA—AGRIMONY.

A. *Eupatoria.* Smith, E. B. vol. ii. p. 346. Sowerby, E. B. vol. xix. pl. 1335.

By road-sides, common.

AGROSTÆMMA—COCKLE.

A. Githago. Smith, E. B. vol. ii. p. 325. Sowerby, E. B. vol. xi. pl. 741.

In corn-fields, frequent.

ALCHEMILLA—LADY'S MANTLE.

A. vulgaris. Smith, E. B. vol. i. p. 223. Sowerby, E. B. vol. ix. pl. 597.

I insert this plant on the authority of Camden's Britannia, where, under the name of "*Stellaria vel Sanicula major,*" it is mentioned as growing at Whiteparish, a very probable situation.

ALISMA—WATER PLANTAIN.

A. ranunculoides. (Lesser water plaintain.) Smith, E. B. vol. ii. p. 205. Sowerby, E. B. vol. v. pl. 326.

Bogs on Alderbury Common; ditches about Downton, abundant.

ANAGALLIS—PIMPERNEL.

A. tenella. Smith, E. B. vol. i. p. 28. Sowerby, E. B. vol. viii. pl. 530.

Bogs on Alderbury common.

ANEMONE.

A. nemorosa flore purpureo. (Wood anemone, with purple flowers.) Smith, E. B. vol. iii. p. 36. Sowerby, E. B. vol. v. pl. 355.

This remarkable variety I have seen but once (and then only sparingly), in a coppice, near the foot of Alderbury Hill.

ANTHRISCUS.

A. vulgaris. (Rough chervil.) Smith, E. B. vol. ii. p. 45. Sowerby, E. B. vol. xii. pl. 818.

Among rubbish on Milford Hill. This species is the *Scandix Anthriscus* of Linneus.

ANTHYLLIS.

A. vulneraria. (Kidney-vetch, or lady's-finger.) Smith, E. B. vol. iii. p. 269. Sowerby, E. B. vol. ii. pl. 104.

This plant seems to be peculiar to chalky soils. It is more abundant on the hills immediately around Salisbury than I have ever seen it elsewhere.

ANTIRRHINUM—SNAP-DRAGON.

A. Cymbalaria. (Ivy-leaved snap-dragon.) Smith, E. B. vol. iii. p. 131. Sowerby, E. B. vol. vii. pl. 502.

On a wall at Laverstock. [DR. SMITH, Fl. Sarisb.]

A. Elatine. (Sharp-pointed fluellin.) Smith, E. B. vol. iii. p. 132. Sowerby, E. B. vol. x. pl. 692.

Corn-fields near the gate at Pitton, abundant; also, between Downton and Redlynch.

A. Orontium. (Calf's snout). Smith, E. B. vol. iii. p. 136. Sowerby, E. B. vol. xvii. pl. 1155.

Corn-fields in chalky soils about Downton.

A. spurium. (Round-leaved fluellin.) Smith, E. B. vol. iii. p. 131. Sowerby, E. B. vol. x. pl. 691.

Corn-fields about Downton; near Clarendon. [DR. SMITH, Fl. Sarisb.]

APIUM.

A. graveolens. (Wild celery.) Smith, E. B. vol. ii. p. 76. Sowerby, E. B. vol. xvii. pl. 1210.

Ditches about Stratford, plentiful.

AQUILEGIA—COLUMBINE.

A. vulgaris. Smith, E. B. vol. iii. p. 33. Sowerby, E. B. vol. v. pl. 297.

Clarendon Wood. I have seen the variety *flore albo* of this plant in one or two spots where the soil was chalky.

ARABIS—WALL CRESS.

A. hirsuta. Smith, E. B. vol. iii. p. 213. Sowerby, E. B. vol. ix. pl. 587.

Turritis hirsuta of Linn.

On the old walls of Clarendon Palace. Found by MR. TURNER on the walls of Old Sarum.

ARENARIA—SANDWORT.

A. rubra. (Purple sandwort.) Smith, E. B. vol. ii. p. 311. Sowerby, E. B. vol. xii. pl. 852.

Sandy parts of Alderbury Common.

A. trinervis. (Plantain-leaved sandwort.) Smith, E. B. vol. xxi. p. 1483. Sowerby, E. B. vol. xxi. pl. 1483.

Under moist hedges near Laverstock.

ARUM—CUCKOW-PINT.

A. maculatum. Smith, E. B. vol. iv. p. 146. Sowerby, E. B. vol. xix. pl. 1298.

Hedge-banks, not uncommon.

ARUNDO—REED.

A. Phragmites. Smith, E. B. vol. i. p. 168. Sowerby, E. B. vol. vi. pl. 401.

Ditches, near Stratford.

ASARUM.

A. Europœum. Smith, E. B. vol. ii. p. 342. Sowerby, E. B. vol. xvi. pl. 1083.

This rare plant, which is described by Sir James Smith as being exclusively a native of the *northern* parts of our island, is to be found under a hedge on the right hand side of a road leading from Redlynch towards Standlynch, just beyond a large chalk pit. It was seen by me there first in the year 1833.

ASPERUGO.

A. procumbens. Smith, E. B. vol. i. p. 265. Sowerby, E. B. vol. x. pl. 661.

This little plant is pronounced in Smith's English Botany to be "*rare.*" On the hills above Wick it is abundant.

ASPERULA—WOODROFFE.

A. cynanchica. (Quinsey wort.) Smith, E. B. vol. i. p. 198. Sowerby, E. B. vol. i. pl. 33.

On the ramparts of Old Sarum.

A. odorata. (Sweet woodrooffe.) Smith, E. B. vol. i. p. 197. Sowerby, E. B. vol. xi. pl. 755.

Clarendon Wood.

ASPIDIUM—SHIELD FERN.

A. Filix mas. Smith, E. B. vol. iv. p. 288. Sowerby, E. B. vol. xxi. pl. 1458.

Alderbury Hill.

ASPLENIUM—SPLEEN WORT.

A. Ruta muraria. Smith, E. B. vol. iv. p. 309. Sowerby, E. B. vol. iii. pl. 150.

Common on walls.

A. Trichomanes. Smith, E. B. vol. iv. p. 305. Sowerby, E. B. vol. viii. pl. 576.

Walls of Salisbury Cathedral.

ASTRAGALUS.

A. glycyphyllos. (Wild liquorice.) Smith, E. B. vol. iii. p. 294. Sowerby, E. B. vol. iii. pl. 203.

Clarendon Wood. This plant grows also, and very luxuriantly, near the great cedar in Farley parish.

ATROPA.

A. Belladonna. (Deadly night-shade.) Smith, E. B. vol. i. p. 317. Sowerby, E. B. vol. ix. pl. 592.

In a lane at the foot of Alderbury Hill.

AURICULARIA.

A. ferruginea. Sowerby, Fungi, vol. i. pl. 26.

Old gate-posts, Clarendon.

A. phosphorea. Sowerby, Fungi, vol. iii. pl. 350.

On rotten wood in a garden at Salisbury.

AVENA—OAT.

A. pubescens. Smith, E. B. vol. i. p. 164. Sowerby, E. B. vol. xxiii. pl. 1640.

High chalky pastures about Downton.

BARBAREA.

B. vulgaris. (Yellow rocket.) Smith, E. B. vol. iii. p. 198. Sowerby, E. B. vol. vii. pl. 443.

Sides of a rivulet in Clarendon Wood. This is the *Erysimum Barbarea* of Linnæus.

BARTRAMIA.

B. pomiformis. Hooker and Taylor, Musc. Brit. p. 144, tab. 23. Sowerby, E. B. vol. xiv. pl. 998.

Bryum pomiforme. Linn.

In dry parts of Alderbury Common.

BARTSIA.

B. Odentites. Smith, E. B. vol. iii. p. 110. Sowerby, E. B. vol. xx. pl. 1415.

Rather common in pastures, in wet situations. This is the *Euphrasia Odentites* of Linnæus.

BERBERIS—BARBERRY.

B. vulgaris. Smith, E. B. vol. ii. p. 184. Sowerby, E. B. vol. i. pl. 49.

Side of a lane leading from Old Sarum to Stratford.

At the time when I collected specimens from these bushes, I was informed that the farmer who rented the land intended to root them out, being persuaded that they occasioned blight in the corn grown near them. This is a pretty general (but, as I have reason to believe, from repeated observations, *unfounded*) notion among agricultural people.

BETONICA—BETONY.

B. officinalis. Smith, E. B. vol. iii. p. 97. Sowerby, E. B. vol. xvi. pl. 1142.

In hedge-rows.

BIDENS—BUR-MARIGOLD.

B. cernua. Smith, E. B. vol. iii. p. 399. Sowerby, E. B. vol. xvi. pl. 1114.

Ditches, near Laverstock Mill.

B. tripartita. Smith, E. B. vol. iii. p. 398. Sowerby, E. B. vol. xvi. pl. 1113.

In Gravel Lane, Downton, abundant.

BLECHNUM.

B. boreale. Smith, E. B. vol. iv. p. 316. Sowerby, E. B. vol. xvii. pl. 1159.

Ditches, dry during the summer, Alderbury Common.

BORVERA.

B. ciliaris. Acharii Lich. Univ. p. 496. Sowerby, E. B. vol. xix. pl. 1352.

Trunks of trees, about Milford. This is *Lichen ciliaris* of Linn.

BOTRYCHIUM.

B. Lunaria. Smith, E. B. vol. iv. p. 328. Sowerby, E. B. vol. v. pl. 318.

Osmunda Lunaria. Linn.

Alderbury Common, near the spot where *Osmunda regalis* is to be found, but in a more moist soil.

BRYUM.

B. argenteum. Linn. (Silvery thread moss.) Hooker and Taylor, Musc. Brit. p. 199, tab. 29. Sowerby, E. B. vol. xxiii. pl. 1602.

On old walls.

BUPLEURUM.

B. rotundifolium. (Thorow-wax.) Smith, E. B. vol. ii. p. 93. Sowerby, E. B. vol. ii. pl. 99.

Corn-fields near Pitton. Very plentiful near Shrewton.

BUTOMUS.

B. umbellatus. (Flowering rush.) Smith, E. B. vol. ii. p. 245. Sowerby, E. B. vol. x. pl. 651.

Rivulets near Stratford and Downton.

CALTHA—MARSH MARIGOLD.

C. palustris. Smith, E. B. vol. iii. p. 59. Sowerby, E. B. vol. viii. pl. 506.

In marshes, common.

CAMPANULA—BELL-FLOWER.

C. glomerata. (Clustered bell-flower.) Smith, E. B. vol. i. p. 292. Sowerby, E. B. vol. ii. pl. 90.

Chalky pastures (in elevated situations), very frequent.

C. hybrida. (Corn bell-flower.) Smith, E. B. vol. i. p. 293. Sowerby, E. B. vol. vi. pl. 375.

Corn-fields near Fisherton. Near the site of Old Sarum. [TURNER and SOWERBY.]

C. Trachelium. (Nettle-leaved bell-flower.) Smith, E. B. vol. i. p. 292. Sowerby, E. B. vol. i. pl. 12.

Clarendon Wood. I have now and then seen, in chalky soils, a variety with *white* flowers. Bell-Mount is one of these situations.

CARDAMINE—LADY'S-SMOCK.

C. *amara.* (Bitter lady's smock.) Smith, E. B. vol. iii. p. 191. Sowerby, E. B. vol. xiv. pl. 1000.

Sides of rivulets, near Fisherton Mill.

CARDUUS—THISTLE.

C. marianus. (Milk thistle.) Smith, E. B. vol. iii. p. 386. Sowerby, E. B. vol. xiv. pl. 976.

Milford Hill, not unfrequent.

CENOMYCE.

C. coccifera. (Scarlet cup lichen.) Acharii Lich. Univ. p. 537, tab. 11, f. 3, quoad apothecam. Sowerby, vol. xxix. pl. 2051.

Mud-walls, about Milford.

This is *Lichen cocciferus* of Linnæus.

C. pyxidata. (Common cup lichen.) Acharii Lich. Univ. p. 534. Sowerby, vol. xx. pl. 1393.

Heath, near Downton.

This is *lichen pyxidatus.* Linn.

C. rangiferina. (Rein-deer moss.) Acharii Lich. Univ. p. 564. Sowerby, E. B. vol. iii. pl. 173.

Near Downton. [Dr. Dickson.]

This is *lichen rangiferinus.* Linn.

CETRARIA.

C. glauca. Acharii Lich. Univ. p. 509. Sowerby, vol. xxiii. pl. 1606.

This species was identified by Dr. Dickson, from specimens collected by Mr. Rooke, from trunks of trees near Downton.

Lichen glaucus of Linn.

CHELIDONIUM—CELANDINE.

C. majus. (Greater celandine.) Smith, E. B. vol. iii. p. 4. Sowerby, E. B. vol. xxii. pl. 1581.

Road-sides, usually among rubbish, about Salisbury.

CHLORA—YELLOW WORT.

C. perfoliata. Smith, E. B. vol. ii. p. 218. Sowerby, E. B. vol. i. pl. 60.

On the ramparts of Old Sarum.

CHRYSANTHEMUM.

C. Leucanthemum. (Ox-eye.) Smith, E. B. vol. iii. p. 449. Sowerby, E. B. vol. ix. pl. 601.

Borders of corn-fields, about Salisbury.

C. segetum. (Corn-marigold.) Smith, E. B. vol. iii. p. 449. Sowerby, E. B. vol. viii. pl. 540.

Corn-fields, not unfrequent. Very abundant about a mile eastward of Salisbury.

CHRYSOSPLENIUM—GOLDEN SAXIFRAGE.

C. oppositifolium. Smith, E. B. vol. ii. p. 260. Sowerby, E. B. vol. vii. pl. 490.

Moist banks, about Alderbury.

CICHORIUM—SUCCORY.

C. Intybus. Smith, E. B. vol. iii. p. 379. Sowerby, E. B. vol. viii. pl. 539.

By road-sides, not uncommon.

CINERARIA—FLEA-WORT.

C. integrifolia. Smith, E. B. vol. iii. p. 444. Sowerby, E. B. vol. iii. pl. 152.

Dry, chalky hills, near Winterslow. [ROBERT WRAY, ESQ.]

CIRCÆA—ENCHANTER'S NIGHTSHADE.

C. lutetiana. (Enchanter's nightshade.) Smith, E. B. vol. i. p. 14. Sowerby, E. B. vol. xv. pl. 1056.

Moist Woods, Alderbury, Hale, &c. Gravel Lane, Downton.

CISTUS.

C. Helianthemum. Smith, E. B. vol. iii. p. 26. Sowerby, E. B. vol. xix. pl. 1321.

On the hills between Wick and Charlton Farm.

CLAVARIA.

C. Coralloides. Sowerby, Fungi, vol. iii. pl. 278.

In the neighbourhood of Downton. [DR. DICKSON.]

C. Hypoxylon. Sowerby, Fungi, vol. i. pl. 55.

On sticks, about Alderbury.

C. ophioglossoides. Linn. à Gm. Sowerby, Fungi, vol. i. pl. 83.

Near Stonehenge. [SIR THOMAS GERY CULLUM, BART.]

CLEMATIS.

C. Vitalba. (Traveller's joy.) Smith, E. B. vol. iii. p. 38. Sowerby, E. B. vol. ix. pl. 612.

In almost every hedge, in this district.

CLINOPODIUM.

C. vulgare. (Wild basil.) Smith, E. B. vol. iii. p. 105. Sowerby, E. B. vol. xx. pl. 1401.

In hedge-rows, abundantly.

CNICUS.

C. eriophorus. (Woolly-headed plume-thistle.) Smith, E. B. vol. iii. p. 390. Sowerby, E. B. vol. vi. pl. 386.

Road-sides, near Clarendon Park.

COLLEMA.

C. nigrescens. Acharii Lich. Univ. p. 646. Sowerby, vol. v. pl. 345.

Trunks of trees, near Stratford. This is *Lichen nigrescens* of Linnæus, Suppl. Plant.

COMARUM—MARSH-LOCKS.

C. palustre. Smith, E. B. vol. ii. p. 433. Sowerby, E. B. vol. iii. pl. 172.

Bogs, Alderbury Common.

CONFERVA.

C. velutina. Sowerby, E. B. vol. xxii. pl. 1556.

This plant, which resembles in colour rich purple velvet, grows on rotten wood at Clarendon.

CONIUM—HEMLOCK.

C. maculatum. Smith, E. B. vol. ii. p. 65. Sowerby, E. B. vol. xvii. pl. 1191.

Hedge-rows, in damp situations about Milford Hill.

CONVALLARIA.

C. majalis. (Lily of the valley.) Smith, E. B. vol. ii. p. 155. Sowerby, E. B. vol. xv. pl. 1035.

Gravely woods. [Mr. Roberts, A.L.S.]

C. multiflora. (Solomon's seal.) Smith, E. B. vol. ii. p. 156. Sowerby, E. B. vol. iv. pl. 279.

Coppice near Whaddon. This is probably the identical spot alluded to by RAY, " a close belonging to the Parsonage of Alderbury," which is mentioned on the authority of Parkinson, (in Gibson's edition of Camden), p. 699.

CONVOLVULUS—BINDWEED.

C. sepium. Smith, E. B. vol. i. p. 284. Sowerby, E. B. vol. v. pl. 313.

Hedges about Downton, frequent.

CONYZA.

C. squarrosa. (Ploughman's spikenard.) Smith, E. B. vol. iii. p. 420. Sowerby, E. B. vol. xvii. pl. 1195.

In poor, stony soils; more frequently about Clarendon than I have seen it elsewhere in the vicinity of Salisbury.

CORNUS—CORNEL.

C. sanguinea. (Common dog-wood.) Smith, E. B. vol. i. p. 221. Sowerby, E. B. vol. iv. pl. 249.

Hedges, on the hill above Wick.

COTYLEDON.

C. Umbilicus. (Navel-wort.) Smith, E. B. vol. ii. p. 314. Sowerby, E. B. vol. v. pl. 325.

Old stone walls, about Dinton.

CUSCUTA—DODDER.

C. Epithymum. (Lesser dodder.) Smith, E. B. vol. ii. p. 25. Sowerby, E. B. vol. i. pl. 55.

On furze, Alderbury Common.

DAPHNE.

D. Laureola. (Spurge laurel.) Smith, E. B. vol. ii. p. 229. Sowerby, E. B. vol. ii. pl. 119.

Plantations, near Winterslow.

DIANTHUS—PINK.

D. Armeria. Smith, E. B. vol. ii. p. 286. Sowerby, E. B. vol. v. pl. 317.

Woods, near Pitton.

DICRANUM.

D. bryoides. Hooker and Taylor, Musc. Brit. p. 88, tab. 16.

Bryum viridulum. Linn.

Moist banks, near Downton.

D. glaucum. Hooker and Taylor, Musc. Brit. p. 92, tab. 16.

Bryum glaucum. Linn.

Heath, near Downton. [Dr. Dickson.]

D. scoparium. Hooker and Taylor, Musc. Brit. p. 101, tab. 18.

Bryum scoparium. Linn.

Hedge-rows, near Downton. [Dr. Dickson.]

D. taxifolium. Taylor and Hooker, Musc. Brit. p. 91, tab. 16.

Hypnum taxifolium. Linn.

Ditch-banks, about Salisbury.

DIGITALIS—FOXGLOVE.

D. purpurea. (Purple foxglove.) Smith, E. B. vol. iii. p. 140. Sowerby, E. B. vol. xix. pl. 1297.

The corolla of this plant is sometimes *white*, in chalky soils, in this country.

It grows most luxuriantly in birch woods near Winterslow, and by the sides of the hill leading from Downton to Redlynch.

DIPSACUS—TEASEL.

D. pilosus. (Lesser teasel.) Smith, E. B. vol. i. p. 193. Sowerby, E. B. vol. xiii. pl. 877.

Ditch-banks, near the Mill, at Milford. Very luxuriantly by the sides of a stream, to the right of the road from Downton to Hale House.

DROSERA—SUN-DEW.

D. longifolia. (Long-leaved sun-dew.) Smith, E. B. vol. ii. p. 123. Sowerby, E. B. vol. xiii. pl. 868.

D. rotundifolia. (Round-leaved sun-dew.) Smith, E. B. vol. ii. p. 122. Sowerby, E. B. vol. xiii. pl. 867.

Both these species are found on bogs near Pitton.

ECHIUM—VIPER-GRASS.

E. vulgare. Smith, E. B. vol. i. p. 268. Sowerby, E. B. vol. iii. pl. 181.

Corn-fields, not unfrequent, especially about Alderbury.

ELEOCHARIS—SPIKE-RUSH.

E. palustris. Smith, E. B. vol. i. p. 62. Sowerby, E. B. vol. ii. pl. 131.

In bogs, on Alderbury Common. This is the *Scirpus palustris* of Linnæus.

ENTEROMORPHA.

E. intestinalis. Greville, Alg. Brit. p. 179.
Ulva intestinalis. Linn.

In rivulets, between Downton and Standlynch Mill.

EPILOBIUM—WILLOW-HERB.

E. hirsutum. Smith, E. B. vol. ii. p. 213. Sowerby, E. B. vol. xii. pl. 838.

River-sides. This plant is, in some places, called *codlings and cream,* an appellation remarkably warranted by the odour which it imparts to the hand when drawn over the backs of the leaves, but this is not always observable in older specimens.

E. montanum. Smith, E. B. vol. ii. p. 214. Sowerby, E. B. vol. xvii. pl. 1177.

Sides of a chalk-pit, near Milford.

E. parviflorum. Smith, E. B. vol. ii. p. 214. Sowerby, E. B. vol. xii. pl. 795.

Sides of rivulets, about Downton.

E. palustre. Smith, E. B. vol. ii. p. 216. Sowerby, E. B. vol. v. pl. 346.

Bogs, on Alderbury Common.

EPIPACTIS—HELLEBORINE.

E. grandiflora. Smith, E. B. vol. iv. p. 43. Sowerby, E. B vol. iv. pl. 271.

Serapias grandiflora. Linn.

Abundantly in the hanging plantations, near Winterslow. I have seen a few specimens in the plantations near Trafalgar Park.

ERICA—HEATH.

E. cinerea. Smith, E. B. vol. ii. p. 226. Sowerby, E. B. vol. xv. pl. 1015.

Alderbury Common.

E. Tetralix. Smith, E. B. vol. ii. p. 226. Sowerby, E. B. vol. xv. pl. 1014.

Grows in the same situation as the preceding species, and mixed with it. There is a variety with a *white* flower, but rare on this common.

ERIOERON.

E. acre. (Flea-bane.) Smith, E. B. vol. iii. p. 422. Sowerby, E. B. vol. xvii. pl. 1158.

In some parts of Clarendon Wood. Also on the Downs between the turnpike-road and Winterslow.

ERIOPHORUM—COTTON-GRASS.

E. polystachion. Smith, E. B. vol. i. p. 67. Sowerby, E. B. vol. viii. p. 563.

E. vaginatum. Smith, E. B. vol. i. p. 66. Sowerby, E. B. vol. xiii. p. 873.

Bogs, on Alderbury Common. The latter species, however, grows very sparingly, and is rather a rare English plant.

ERODIUM—STORK'S-BILL.

E. cicutarium. Smith, E. B. vol. iii. p. 229. Sowerby, E. B. vol. xxv. pl. 1768.

In sandy parts of Clarendon Wood, and of Alderbury Common.

ERVUM—TARE.

E. hirsutum—(Hairy tare.) Smith, E. B. vol. iii. p. 289. Sowerby, E. B. vol. xiv. pl. 970.

Corn-fields, near Redlynch, and near Bemerton.

E. tetraspermum—(Four-seeded tare.) Smith, E. B. vol. iii. p. 288. Sowerby, E. B. vol. xvii. pl. 1223.

Woods, near Winterslow.

ERYTHRÆA.

E. centaurium. (Centaury.) Smith, E. B. vol. i. p. 320. Sowerby, E. B. vol. vi. pl. 417.

Clarendon Wood. But it is not very uncommon in other dry woods and pastures in this district.

EUONYMUS—SPINDLE-TREE.

E. europæus. Smith, E. B. vol. i. p. 329. Sowerby, E. B. vol. vi. p. 362.

This handsome shrub is not unfrequent in hedges, on a chalky soil, about Salisbury and Downton.

EUPATORIUM.

E. cannabinum. Smith, E. B. vol. iii. p. 400. Sowerby, E. B. vol. vi. pl. 428.

On the banks of rivers, common.

EUPHORBIA—SPURGE.

E. amygdaloides. Smith, E. B. vol. iv. p. 67. Sowerby, E. B. vol. iv. pl. 256.

In a wood near Dinton.

E. exigua. (Dwarf spurge.) Smith, E. B. vol. iv. p. 60. Sowerby, E. B. vol. xix. pl. 1336.

Corn-fields, bordering on Wykedowns. This species is mentioned in Smith's "Compendium Fl. Brit." as being rare.

EUPHRASIA—EYE-BRIGHT.

E. officinalis. Smith, E. B. vol. iii. p. 122. Sowerby, E. B. vol. xx. pl. 1416.

This elegant little plant is one of the most abundant of our chalk plants, in elevated situations.

FAGUS—BEECH.

F. sylvatica. Smith, E. B. vol. iv. p. 152. Sowerby, E. B. vol. xxvi. pl. 1846.

Woods, about Winterslow.

FEDIA.

F. olitoria. (Corn salad.) Smith, E. B. vol. i. p. 45. Sowerby, E. B. vol. xii. pl. 811.

This is the *Valeriana Locusta* of Linnæus.

In corn-fields, near the foot of Alderbury Hill.

FIBRILLARIA.

F. vinaria. Sowerby, fungi, pl. 432.

This fungus is seen hanging from the roofs of wine-vaults, whence it has its trivial name. In a wine-vault at Salisbury I have collected specimens a foot and a-half in length.

FRAGARIA—STRAWBERRY.

F. vesca. (Wood-strawberry.) Smith, E. B. vol. ii. p. 414. Sowerby, E. B. vol. xxii. pl. 1524.

Clarendon Wood.

FRAXINUS—ASH.

F. excelsior. Smith, E. B. vol. i. p. 13. Sowerby, E. B. vol. xxiv. pl. 1692.

Hedge-rows, near Laverstock, and about Downton.

FUNARIA.

F. hygrometrica. Taylor and Hooker, Musc. Brit. p. 121, tab. 20.

Mnium hygrometricum. Linn.

Near Downton. [DR. DICKSON.]

GALEOBDOLEN.

G. luteum. (Yellow dead-nettle.) Smith, E. B. vol. iii. p. 96. Sowerby, E. B. vol. xi. pl. 787.

Clarendon Wood; and in hedge-rows, near Downton.

GALEOPSIS—HEMP-NETTLE.

G. Ladanum. (Red Hemp-nettle.) Smith, E. B. vol. iii. p. 93. Sowerby, E. B. vol. xiii. pl. 884.

Corn-fields, near Pitton.

G. Tetrahit. (Hemp-nettle.) Smith, E. B. vol. iii. p. 94. Sowerby, E. B. vol. iii. pl. 207.

Coppice, near Alderbury; hedge-rows, near Downton; and, abundantly, by the side of the path leading from the bridge, near Fisherton Mill, to West Harnham Mill.

GALIUM.

G. Mollugo. Smith, E. B. vol. i. p. 208. Sowerby, E. B. vol. xxiv. pl. 1673.

Hedge-rows, common.

G. palustre. (Marsh goose-grass.) Smith, E. B. vol. i. p. 199. Sowerby, E. B. vol. xxvi. pl. 1857.

Stratford Marsh.

G. tricorne. Smith, E. B. vol. i. p. 205. Sowerby, E. B. vol. xxiii. pl. 1641.

Fields, near Downton.

GENISTA—GREENWEED.

G. anglica. (Needle greenweed.) Smith, E. B. vol. iii. p. 264. Sowerby, E. B. vol. ii. pl. 132.

Alderbury Common.

G. tinctoria. (Dyer's greenweed.) Smith, E. B. vol. iii. p. 263. Sowerby, E. B. vol. i. pl. 44.

Wood, near West Deane, abundantly.

GENTIANA—GENTIAN.

G. amarella. Smith, E. B. vol. ii. p. 30. Sowerby, E. B. vol. iv. pl. 236.

Chalk Hills, near Clarendon; also, near Winterslow.

G. campestris. Smith, E. B. vol. ii. p. 31. Sowerby, E. B. vol. iv. pl. 237.

On the ascent to the ancient camp, called Clerebury, or Clarebury.

GERANIUM—CRANE'S-BILL.

G. columbinum. Smith, E. B. vol. iii. p. 241. Sowerby, E. B. vol. iv. pl. 259.

One of the least common species of *geranium* in this country. I have found it about Pitton.

G. dissectum. Smith, E. B. vol. iii. p. 241. Sowerby, E. B. vol. xi. pl. 753.

Dry banks, in chalky places, near the three gates on the Salisbury road from Downton; also about Pitton.

G. phœum. Smith, E. B. vol. iii. p. 232. Sowerby, E. B. vol. v. pl. 322.

Just within the gate (called *Eyre's Gutter Gate*) of a meadow between Alderbury and Standlynch. It was first pointed out to me by *Mr. Roberts, A.L.S.* I have never found it in any other part of England, except in the grounds of Sir Thomas Slingsby, Bart. at Scriven, in Yorkshire. It is one of the *plantæ rariores* of England.

G. pratense. Smith, E. B. vol. iii. p. 235. Sowerby, E. B. vol. vi. pl. 404.

This species, one of the most elegant of our British plants, I have not seen growing otherwise than very sparingly in Wiltshire. I have found it near the grove beyond Fisherton churchyard, and in a hedge-row near the south-west corner of Downton.

GEUM—AVENS.

G. rivale. (Water avens.) Smith, E. B. vol. ii. p. 431. Sowerby, E. B. vol. ii. pl. 106.

Stratford Marsh.

GLECHOMA.

G. hederacea. (Ground ivy.) Smith, E. B. vol. iii. p. 88. Sowerby, E. B. vol. xii. pl. 853.

On hedge-banks, not uncommon.

GLYCERIA—SWEET-GRASS.

G. rigida. Smith, E. B. vol. i. p. 119. Sowerby, E. B. vol. xx. pl. 1371.

Road-sides, between Milford and Clarendon.

This is the *Poa rigida* of Linnæus.

GRAPHIS.

G. scripta. Acharii Lich. Univ. p. 265.

Lichen scriptus. Linn.

Trunks of birch trees, in Clarendon Park.

GRIMMEA.

G. pulvinata. Hooker and Taylor, p. 68, tab. 13. Sowerby, E. B. vol. xxiv. pl. 1728.

Bryum pulvinatum. Linn.

On old walls, frequent.

GYMNOSTOMUM.

G. pyriforme. Hooker and Taylor, Musc. Brit. p. 24, tab. 7. Sowerby, E. B. vol. vi. pl. 412.

Moist banks, not far from Fovant.

G. truncatulum. Hooker and Taylor, Musc. Brit. p. 22. tab. 7. Sowerby, E. B. vol. xxviii. pl. 1975.

Banks, near Britford.

HEDYSARUM—SAINT-FOIN.

H. Onobrychis. Smith, E. B. vol. iii. p. 292. Sowerby, E. B. vol. ii. pl. 96.

Borders of corn-fields, near Mr. Wyndham's grounds, Salisbury.

HELLEBORUS—HELLEBORE.

H. fœtidus. (Fœtid hellebore.) Smith, E. B. vol. iii. p. 58. Sowerby, E. B. vol. ix. pl. 613.

Clarendon Wood.

H. viridis. (Green hellebore.) Smith, E. B. vol. iii. p. 57. Sowerby, E. B. vol. iii. pl. 200.

On the Borders of Clarendon Wood. [Dr. Smith.]

HIERACIUM—HAWKWEED.

H. murorum. (Wall hawkweed.) Smith, E. B. vol. iii. p. 359. Sowerby, E. B. vol. xxix. pl. 2082.

Walls of the Close of Salisbury.

H. subaudum. Smith, E. B. vol. iii. p. 367. Sowerby, E. B. vol. v. pl. 349.

Coppice at the foot of Alderbury Hill.

HIPPOCREPIS—HORSESHOE-VETCH.

H. comosa. (Hairy horseshoe vetch.) Smith, E. B. vol. iii. p. 291. Sowerby, E. B. vol. i. pl. 31.

This plant is almost peculiar to chalky soils, but is not frequent in the vicinity of Salisbury. I have seen it only on the hills near Fisherton.

HUMULUS—HOP.

H. Lupulus. Smith, E. B. vol. iv. p. 240. Sowerby, E. B. vol. vi. pl. 427.

Hedge-rows, near Alderbury, and on the road-sides between Salisbury and Downton.

HYDROCOTYLE.

H. vulgaris. (White-rot.) Smith, E. B. vol. ii. p. 96. Sowerby, E. B. vol. xi. pl. 751.

Rivulets on Alderbury Common.

HYOSCYAMUS—HENBANE.

H. niger. Smith, E. B. vol. i. p. 315. Sowerby, E. B. vol. ix. pl. 591.

Dr. Smith, in his "*Flora Sarisburiensis,*" mentions his having found this plant in a gravelly soil, at Alderbury, and in a chalky loam at Stratford and Milford.

HYPERICUM.

H. Androsæmum. (Tatsan.) Smith, E. B. vol. iii. p. 323. Sowerby, E. B. vol. xviii. pl. 1225.

Hedges, about half a mile distant from Downton, on the road to Salisbury. Clarendon Wood. [Dr. H. Smith.]

H. elodes. Smith, E. B. vol. iii. p. 330. Sowerby, E. B. vol. ii. pl. 109.

Bogs, on Alderbury Common.

H. humifusum. (Creeping hypericum.) Smith, E. B. vol. iii. p. 326. Sowerby, E. B. vol. xviii. pl. 1226.

Boggy ground, at West Dean.

H. perforatum. Smith, E. B. vol. iii. p. 325. Sowerby, E. B. vol. v. pl. 295.

Under hedges, common.

H. pulchrum. Smith, E. B. vol. iii. p. 329. Sowerby, E. B. vol. xviii. pl. 1227.

Not unfrequent about Dinton,—a most elegant plant.

H. quadrangulum. (Square-stem'd hypericum.) Smith, E. B. vol. iii. p. 324. Sowerby, E. B. vol. vi. pl. 370.

River-banks, at West-Harnham. Luxuriantly by the sides of a rivulet, west of Standlynch Mill.

HYPNUM.

H. complanatum. Hooker and Taylor, Musc. Brit. p. 152, tab. 24. Sowerby, E. B. vol. xxi. pl. 1492.

Trunks of trees.

H. confertum. Hooker and Taylor, Musc. Brit. p. 178. tab. 26. Sowerby, E. B. vol. xxxiv. pl. 2407.

On banks.

H. cupressiforme. Hooker and Taylor, Musc. Brit. p. 189, tab. 27. Sowerby, E. B. vol. xxvi. pl. 1860.

On banks.

H. curvatum. Hooker and Taylor, Musc. Brit. p. 169, tab. 25. Sowerby, E. B. vol. xxii. pl. 1566.

On trunks of trees, near Downton. [Dr. Dickson.]

H. filicinum. Hooker and Taylor, Musc. Brit. p. 183, tab. 26. Sowerby, E. B. vol. xxii. pl. 1570.

West Dean Wood.

H. lutescens. Hooker and Taylor, Musc. Brit. p. 166, tab. 25. Sowerby, E. B. vol. xix. pl. 1301.

On dry banks, near Downton. [Dr. Dickson.]

H. myosuroides. Hooker and Taylor, Musc. Brit. p. 169, tab. 25. Sowerby, E. B. vol. xxii. pl. 1567.

Trunks of trees, about Clarendon Park.

H. prælongum. Hooker and Taylor, Musc. Brit. p. 172, tab. 25. Sowerby, E. B. vol. xxix. pl. 2035.

Shady banks, near Downton. [Dr. Dickson.]

H. proliferum. Hooker and Taylor, Musc. Brit. p. 170, tab. 25. Sowerby, E. B. vol. xxi. pl. 1494.

Clarendon Wood.

H. Rutabulum. Hooker and Taylor, Musc. Brit. p. 176, tab. 26.

Moist banks, frequent.

H. Schreberi. Hooker and Taylor, Musc. Brit. p. 159, tab. 24. Sowerby, E. B. vol. xxiii. pl. 1621.

Banks about Downton. [DR. DICKSON.]

H. sericeum. Hooker and Taylor, Musc. Brit. p. 165, tab. 25. Sowerby, E. B. vol. xxi. pl. 1445.

Not uncommon on trunks of trees.

H. serpens. Hooker and Taylor, Musc. Brit. p. 155, tab. 24. Sowerby, E. B. vol. xv. pl. 1037.

Trunks of trees, near Laverstock.

H. scuiroides. Hooker and Taylor, Musc. Brit. p. 112, tab. 20.

Trunks of trees.

H. triquetrum. Hooker and Taylor, Musc. Brit. p. 182. tab. 26. Sowerby, E. B. vol. xxiii. pl. 1662.

Ditch-banks, Alderbury.

H. velutinum. Hooker and Taylor, Musc. Brit. p. 177. tab. 25. Sowerby, E. B. vol. xxii. pl. 1568.

Dry banks, Milford.

ILEX—HOLLY.

I. Aquifolium. Smith, E. B. vol. i. p. 227. Sowerby, E. B. vol. vii. pl. 496.

Clarendon Wood.

IMPATIENS.

I. Noli me tangere. (Touch-me-not.) Smith, E. B. vol. i. p. 299. Sowerby, E. B. vol. xiv. pl. 937.

Seen by me on the sides of the river below Crane bridge, Salisbury. It is possible that the seeds may have been wafted thither from some garden.

INULA.

I. dysenterica. (Common Fleabane.) Smith, E. B. vol. iii. p. 441. Sowerby, E. B. vol. xvi. pl. 1115.

Common in stagnant water.

I. Helenium. (Elecampane.) Smith, E. B. vol. iii. p. 441. Sowerby, E. B. vol. xxii. pl. 1546.

Left bank of the river, near Bemerton; also in meadows near West Harnham mill.

IRIS—FLAG.

I. fœtidissima. (Fetid flag.) Smith, E. B. vol. i. p. 49. Sowerby, E. B. vol. ix. p. 596.

Under hedges, in a lane, near Milford.

I. pseud-acorus. (Yellow flag.) Smith, E. B. vol. i. p. 48. Sowerby, E. B. vol. ix. p. 578.

By river-sides; not uncommon in this county.

JUNIPERUS.—JUNIPER.

J. communis. (Common juniper). Smith, E. B. vol. iv. p. 251. Sowerby, E. B. vol. xvi. pl. 1100.

On the higher parts of Thorney Down.

JURGERMANNIA.

J. complanata. Sowerby, E. B. vol. xxxv. pl. 2499.

Trunks of fir-trees.

J. dilatata. Hooker and Taylor, Musc. Brit. p. 239. Sowerby, E. B. vol. xvi. pl. 1086.

J. platyphylla. Hooker and Taylor, Musc. Brit. p. 237. Sowerby, E. B. vol. xii. pl. 780.

Trunks of trees in Clarendon Wood.

J. tamariscifolia. [Dr. Dickson.]

Trunks of trees.

J. trichomanis. Hooker and Taylor, Musc. Brit. p. 235. Sowerby, E. B. vol. xxvii. pl. 1875.

About Downton. [Dr. Dickson.]

LAMIUM—DEAD-NETTLE.

L. amplexicaule. Smith, F. B. vol. iii. p. 92. Sowerby, E. B. vol. ii. pl. 770.

Corn-fields about Salisbury.

LATHRÆA—TOOTH-WORT.

L. squamaria. Smith, E. B. vol. iii. p. 127. Sowerby, E. B. vol. i. pl. 50.

Plantations near Trafalgar Park.

LATHYRUS—VETCHLING.

L. aphaca. (Yellow Vetchling.) Smith, E. B. vol. iii. p. 274. Sowerby, E. B. vol. xvii. pl. 1167.

Hedge-rows, near Alderbury.

L. Nissolia. (Grass vetchling.) Smith, E. B. vol. iii. p. 275. Sowerby, E. B. vol. ii. pl. 112.

This plant was found by DR. HENRY SMITH (he says) abundantly, in a clayey soil, near Clarendon Wood; also on the borders of corn-fields, near the Devizes road; but I have found it only in one spot (and there very scantily) in Wiltshire, namely, in a small coppice a little beyond the 1-mile stone on the turnpike road from Salisbury to London. It is a very beautiful species.

L. sylvestris. (Wood vetchling.) Smith, E. B. vol. iii. p. 277. Sowerby, E. B. vol. xii. pl. 805.

Between Compton and Dinton.

LECANORA.

L. angulosa. Acharii Lich. Univ. p. 364.

In the clefts of the bark of very old trees.

L. hæmatomma. Acharii Lich. Univ. p. 388. Sowerby, E. B. vol. iv. pl. 223.

Lichen coccineus. Dickson, Cr. Brit.

Stonehenge.

L. saxicola. Acharii Syn. Meth. Lich. p. 180. Sowerby, E. B. vol. xxiv. pl. 1695.

Parmelia saxicola. Acharii Lich. Univ. p. 431.

On small stones, Salisbury Plain. [TURNER and SOWERBY.]

LECIDEA.

L. confluens. Acharii Lich. Univ. p. 174.

Stonehenge. [TURNER and SOWERBY.]

L. Parasoma. Acharii Lich. Univ. p. 175.

Trunks of firs about Salisbury.

LIGUSTRUM—PRIVET.

L. vulgare. Smith, E. B. vol. i. p. 13. Sowerby, E. B. vol. xi. pl. 764.

The chalky soil of the vicinity of Salisbury is very favourable to the growth of this elegant and oriental-like shrub. It is abundant in hedge-rows.

LINUM—FLAX.

L. catharticum. (Cathartic flax.) Smith, E. B. vol. ii. p. 119. Sowerby, E. B. vol. vi. pl. 382.

Chalky fields, near Salisbury, by the sides of the road leading to Devizes; also between Downton and Redlinch.

L. usitatissimum. (Common flax.) Smith, E. B. vol. ii. p. 119. Sowerby, E. B. vol. xix. pl. 1357.

This elegant plant is frequent in corn-fields about Clarendon.

LISTERA.

L. nidus avis. (Bird's nest Ophrys.) Smith, E. B. vol. iv. p. 38. Sowerby, E. B. vol. i. pl. 48.

Ophrys nidus avis. Linn.

Winterslow Woods. Specimens from them were first presented to me by my friend P. B. Brodie, Esq.

L. ovata. (Sway blade.) Smith, E. B. vol. iv. p. 36. Sowerby, E. B. vol. xxii. pl. 1548.

Ophrys ovata. Linn.

In a coppice near Whaddon.

LITHOSPERMUM—GROMWELL.

L. arvense. (Corn Gromwell.) Smith, E. B. vol. i. p. 255. Sowerby, E. B. vol. ii. pl. 123.

In corn-fields, but not very frequent.

L. officinale. (Common Gromwell.) Smith, E. B. vol. i. p. 255. Sowerby, E. B. vol. ii. pl. 134.

Lanes, in a gravelly soil; common.

LONICERA—HONEY-SUCKLE.

L. Periclymenum. Smith, E. B. vol. i. p. 326. Sowerby, E. B. vol. xii. pl. 800.

Frequent, in hedge-rows.

LOTUS.

L. corniculatus. Smith, E. B. vol. iii. p. 312. Sowerby, E. B. vol. xxx. pl. 2090.

Very common in this district, especially among luxuriant herbage.

L. major. Smith, E. B. vol. iii. p. 313. Sowerby, E. B. vol. xxx. pl. 2091.

In wet bushy spots, about Downton.

LYCHNIS—CAMPION.

L. dioica. Smith, E. B. vol. ii. p. 328.

Var. *α.* (*Red* Campion.) Sowerby, E. B. vol. xxii. pl. 1579.
Under hedges, in moist situations, frequent.

Var. *β.* (*White* Campion.) Sowerby, E. B. vol. xxii. pl. 1580.
In fields, frequent.

L. flos cuculi. (Cuckow flower.) Smith, E. B. vol. ii. p. 326. Sowerby, E. B. vol. viii. pl. 573.

In meadows, common.

LYCOPERDON.

L. proteus. Sowerby, Fungi, vol. iii. pl. 332.

In the neighbourhood of Downton. [Dr. Dickson.]

LYCOPODIUM—CLUB MOSS.

L. inundatum. (Moist club moss.) Smith, E. B. vol. iv. p. 332. Sowerby, E. B. vol. iv. pl. 239.

Marshy ground, on Alderbury common.

L. Selago. Smith, E. B. vol. iv. p. 333. Sowerby, vol. iv. pl. 233.

On dry parts of Alderbury common.

LYSIMACHIA—LOOSE STRIFE.

L. nemorum. (Yellow Pimpernel or Wood Loose strife.) Smith, E. B. vol. i. p. 279. Sowerby, E. B. vol. viii pl. 527.
River-sides, Fisherton.

L. Nummularia. (Money-wort.) Smith, E. B. vol. i. p. 279. Sowerby, E. B. vol. viii. pl. 528.
This elegant little creeping plant grows rather plentifully in boggy ground, near West Deane.

L. vulgaris. (Common Loose-strife.) Smith, E. B. vol. i. p. 277. Sowerby, E. B. vol. xi. pl. 761.
River-sides, near West Harnham.

LYTHRUM.

L. Salicaria. Smith, E. B. vol. ii. p. 343. Sowerby, E. B. vol. xv. pl. 1061.

The spots where I have seen this plant most abundant and flourishing are about Stratford.

MALVA—MALLOW.

M. moschata. (Musk mallow.) Smith, E. B. vol. iii. p. 247. Sowerby, E. B. vol. xi. pl. 754.

I have not seen this plant growing anywhere abundantly; but it is to be found, sparingly, in Clarendon wood, and also by the side of a path leading from Downton into the road to Christchurch, close to the border of the county of Wilts. It is remarkable for its elegance both of flower and leaf.

M. rotundifolia. (Round-leaved mallow.) Smith, E. B. vol. iii. p. 246. Sowerby, E. B. vol. xvi. pl. 1092.

Borders of corn-fields, Milford Hill.

M. sylvestris. (Common mallow.) Smith, E. B. vol. iii. p. 245. Sowerby, E. B. vol. x. pl. 671.

The only reason for including *this* species (which is extremely common throughout the southern counties) in the present catalogue is the extraordinary luxuriance of its growth in a chalk pit near Milford; *some* specimens here I have found about ten feet in height.

MARRUBRIEM—(WHITE HOREHOUND.)

M. vulgare. Smith, E. B. vol. iii. p. 103. Sowerby, E. B. vol. vi. pl. 410.

Hedge-rows, near the road from Downton to Salisbury.

MATRICARIA.

M. chamomilla. (Bastard chamomile.) Smith, E. B. vol. iii. p. 454. Sowerby, E. B. vol. xviii. pl. 1232.

This plant, said to possess medicinal properties similar to those of the true chamomile *(anthemis nobilis)*, is not uncommon on the sides of corn-fields, and in rubbish, about Salisbury.

MEDICAGO—MEDIC.

M. lupulina. Smith, E. B. vol. iii. p. 318. Sowerby, E. B. vol. xiv. pl. 971.

Clarendon Wood.

MELAMPYRUM.

M. pratense. Smith, E. B. vol. iii. p. 125. Sowerby, E. B. vol. ii. p. 113.

In shady meadows near Laverstock, but not common.

MENTHA—MINT.

M. hirsuta. (Hairy mint.) Smith, E. B. vol. iii. p. 78. Var. *a*. Sowerby, E. B. vol. vii. pl. 447.

Sides of a path leading from Downton to Charlton.

M. sylvestris. (Horse mint.) Smith, E. B. vol. iii. p. 72. Sowerby, E. B. vol. x. pl. 686.

Moist hedge-rows in the parish of Wick, near Downton.

M. viridis. (Spear mint.) Smith, E. B. vol. iii. p. 75. Sowerby, E. B. vol. xxxiv. pl. 2424.

Stratford Marsh.

MENYANTHES—BUCK-BEAN.

M. *trifoliata.* Smith, E. B. vol. i. p. 274. Sowerby, E. B. vol. vii. pl. 495.

Marshes, Alderbury Common.

MIMULUS—MONKEY-FLOWER.

M. guttatus.

This plant has not hitherto obtained a place in the British Flora (though there is no reason on the score either of climate or locality why it should not inhabit the meadows of Downton as well as those of Virginia), nor ought it, perhaps, to be introduced into this catalogue merely in consequence of my having found only two specimens, which might have sprung from seeds wafted down the Avon from some garden ; but I have considered it right to note down the spot where so handsome a stranger has been seen, in a wild state, that botanists may search it well hereafter, and endeavour to ascertain whether it become ultimately naturalised amongst us. I found it (in company with my friend Robert Wray, Esq. F.L.S.) on the side of a branch of the Avon, at New Farm, near Downton, close to a little wooden bridge, leading from opposite that farm to the meadows, in Sept. 1830, when it was in full flower.

MONOTROPA—YELLOW BIRD'S NEST.

M. Hypopithys. Smith, E. B. vol. ii. p. 249. Sowerby, E. B. vol. i. pl. 69.

This singular parasitic I have found in the woods of Winterslow, but I believe it to be uncommon in the neighbourhood of Salisbury.

MONTIA—BLINKS.

M. fontana. Smith, E. B. vol. i. p. 186. Sowerby, E. B. vol. xvii. pl. 1206.

Bogs, near Alderbury.

MYRRHIS—CICELY.

M. temulenta. Smith, E. B. vol. ii. p. 51. Sowerby, E. B. vol. xxii. pl. 1521.

Hedge-rows about Downton.

This is the *chærophyllum temulentum* of Linnæus.

NARCISSUS.

N. pseudo narcissus. (Daffodil.) Smith, E. B. vol. ii. p. 132. Sowerby, E. B. vol. i. pl. 17.

This plant grew so abundantly in March 1793, in a wood behind Ivychurch, as to render a large space of ground of an *uninterrupted* yellow colour. I have not had an opportunity of visiting the spot at the same season of the year since.

NARTHECIUM—BOG-ASPHODEL.

N. *ossifragum.* Smith, E. B. vol. ii. p. 151. Sowerby, E. B. vol. viii. pl. 535.

Bogs, near Alderbury.

NEOTTIA.

N. spiralis. (Sweet lady's traces.) Smith, E. B. vol. iv. p. 35. Sowerby, E. B. vol. viii. pl. 541.

Ophrys spiralis. Linn.

In August, 1813, I saw this plant literally *covering*, in full flower, a corn-field, at that time fallow, between New Court Farm and Charlton; but I have looked for it, in subsequent years, without success, though it is scarcely to be imagined that the species is entirely extinct there.

Laverstock Down. [DR. H. SMITH.]

NECKERA.

N. crispa. Hooker and Taylor, Musc. Brit. p. 136, tab. 22. Sowerby, E. B. vol. ix. pl. 617.

Hypnum crispum. Linn.

Near Milford, on trunks of trees.

NEPETA—CAT-MINT.

N. cataria. Smith, E. B. vol. iii. p. 70. Sowerby, E. B. vol. ii. pl. 137.

In a lane leading from St. Martin's churchyard, Salisbury, into the turnpike road.

NOSTOC—STAR JELLY.

N. commune. Sowerby, E. B. pl. 4, p. 61.

Grows in great profusion on the acclivity of the hill, near Winterslow.

NUPHAR—YELLOW WATER LILY.

N. lutea. Smith, E. B. vol. iii. p. 15. Sowerby, E. B. vol. iii. pl. 159.

This species (*nymphæa lutea,* of Linnæus) is noted down among my botanical memoranda as having been found by me "near Salisbury," but the exact place is not mentioned, nor do I recollect it now.

NYMPHÆA—WHITE WATER LILY.

N. alba. Smith, E. B. vol. iii. p. 14. Sowerby, E. B. vol. iii. pl. 160.

In the river Avon, near Stratford.

ŒNANTHE—DROP-WORT.

Œ. fistulosa. Smith, E. B. vol. ii. p. 68. Sowerby, E. B. vol. vi. pl. 363.

Marshy ground, near Alderbury.

ONONIS—REST-HARROW.

O. arvensis. Smith, E. B. vol. iii. p. 267. Sowerby, E. B. vol. x. pl. 682.

Very abundant on the borders of corn-fields in all chalky parts of Wiltshire.

I was satisfied many years ago that *O. arvensis,* and *O. spinosa* of Linnæus, were, in reality, the same species, and that the thorns were the result merely of the nature of the soil, or of the stage of growth. The latter opinion was adopted by Sir James Smith, who afterwards expunged *O. spinosa* from the *Flora Britannica,* inserting the words "*demum spinascentibus*" among the characters of *O. arvensis.*

OPHIEGLOSSUM—ADDER'S TONGUE.

O. vulgatum. Smith, E. B. vol. iv. p. 329. Sowerby, E. B. vol. ii. pl. 108.

This species is not properly denominated "*vulgatum,*" so far as relates to England, being far from common amongst us. In Wiltshire it is seen only in moist ground. It grows not far from Whaddon, but not abundantly.

OPHRYS.

O. apifera. (Bee ophrys.) Smith, E. B. vol. iv. p. 30. Sowerby, E. B. vol. vi. pl. 383.

At Old Sarum. Mr. Roberts, A. L. S., found it near Odstock.

ORCHIS.

O. bifolia. (Butterfly orchis.) Smith, E. B. vol. iv. p. 9. Sowerby, E. B. vol. i. pl. 22.

This elegant plant was, at one time, plentiful in a coppice near Whaddon.

O. conopsea. (Aromatic orchis.) Smith, E. B. vol. iv. p. 23. Sowerby, E. B. vol. i. pl. 10.

Meadows between Stratford and Durnford.

O. maculata. (Spotted orchis.) Smith, E. B. vol. iv. p. 22. Sowerby, E. B. vol. ix. pl. 632.

Meadows, near Milford; and at Downton. In a wood, not far from West Deane, I have seen a variety of this species with white flowers and no spots on the leaves.

O. mascula. (Early purple orchis.) Smith, E. B. vol. iv. p. 11. Sowerby, E. B. vol. ix. pl. 631.

In meadows, not uncommon.

O. Morio. (Green-winged meadow orchis.) Smith, E. B. vol. iv. p. 12. Sowerby, E. B. vol. xxix. pl. 2059.

Meadows about Salisbury and Downton.

O. ustulata. (Dwarf orchis.) Smith, E. B. vol. iv. p. 12. Sowerby, E. B. vol. i. pl. 18.

On the chalk hills near Laverstock. [Mr. Lake.]

ORIGANUM—MARJORAM.

O. vulgare. Smith, E. B. vol. iii. p. 106. Sowerby, E. B. vol. xvi. pl. 1143.

Dry banks and borders of corn-fields, in a chalky soil, very frequent.

ORNITHOGALUM—STAR OF BETHLEHEM.

O. pyrenaicum. Smith, E. B. vol. ii. p. 143. Sowerby, E. B. vol. vii. pl. 499.

This beautiful plant grows almost as abundantly as the grass in a small coppice to the left of the road leading from Pitton to Grinstead; but I have never seen it elsewhere in Wiltshire.

ORNITHOPUS—BIRD'S-FOOT.

O. perpusillus. (Diminutive bird's foot.) Smith, E. B. vol. iii. p. 290. Sowerby, E. B. vol. vi. pl. 369.

On dry parts of Alderbury Common, and on the chalky ridge leading from the walls of Mr. Wyndham's grounds, at Salisbury, towards Bishop's Down.

OROBANCHE.

O. major. (Larger broom rape.) Smith, E. B. vol. iii. p. 146. Sowerby, E. B. vol. vi. pl. 421.

Dry banks by the sides of the road from Amesbury to Stonehenge.

OROBUS.

O. tuberosus. Smith, E. B. vol. iii. p. 272. Sowerby, E. B. vol. xvii. pl. 1153.

Wood, near Dinton.

ORTHOTRICHUM.

O. anomalum. Hooker and Taylor, Musc. Brit. p. 126, tab. 21. Sowerby, E. B. vol. xx. pl. 1423.

On walls.

O. diaphanum. Hooker and Taylor, Musc. Brit. p. 128, tab. 21. Sowerby, E. B. vol. xix. pl. 1324.

On trees.

OSMUNDA—OSMUND.

O. regalis. (Osmund royal.) Smith, E. B. vol. iv. p. 327. Sowerby, E. B. vol. iii. pl. 209.

Side of a hedge on Alderbury Common, not far from Clarendon Park.

PAPAVER—POPPY.

P. dubium. (Long smooth-headed poppy.) Smith, E. B. vol. iii. p. x. Sowerby, E. B. vol. ix. pl. 644.

Road-sides, near the gate of the turnpike-road from Salisbury to Alderbury.

P. hybridum. (Hybrid poppy.) Smith, E. B. vol. iii. p. 9. Sowerby, E. B. vol. i. pl. 43.

Border of a corn-field by the side of the road from Milford to Clarendon. I believe this plant to be rare.

P. somniferum. (White poppy.) Smith, E. B. vol. iii. p. 11. Sowerby, E. B. vol. xxx. pl. 2145.

Hedge-rows in a lane leading from Old Sarum to the Avon. It is probable that the seeds may have originally been conveyed hither by accident, from a neighbouring garden. Sir James Smith considers it a doubtful native of our island.

PARMELIA.

P. Borreri. Acharii Lich. Univ. p. 461.

On a heath near Downton [MRS. ROOKE,] identified by DR. DICKSON.

P. olivacea. Acharii Lich. Univ. p. 462.

Trunks of lime trees, in Clarendon Wood.

P. pulverulenta. Acharii Lich. Univ. p. 473.

Bark of lime trees in Clarendon Wood.

PASTINACA—PARSNEP.

P. sativa. (Wild parsnep.) Smith, E. B. vol. ii. p. 101. Sowerby, E. B. vol. viii. pl. 556.

Borders of corn-fields, frequent.

PEDICULARIS—LOUSE-WORT.

P. palustris. (Marsh louse-wort.) Smith, E. B. vol. iii. p. 129. Sowerby, E. B. vol. vi. pl. 399.

Stratford Marsh.

PELTIDEA.

P. canina. Acharii Lich. Univ. p. 517.

Ditch-banks, about Alderbury.

PEZIZA.

P. epidendra. Sowerby, Fungi, vol. i. pl. 13.

This beautiful scarlet fungus I have seen on rotten sticks in the lane crossing the Downton road, by the pales of Lord Radnor's grounds, to Britford.

PHASCUM.

P. cuspidatum. Hooker and Taylor, Musc. Brit. p. 8, tab. 5. Sowerby, E. B. vol. xxix. pl. 2025.

Var. a. apiculatum. Phascum acaulor. Linn.

Sandy banks.

PHLEBIA.

P. merismoides. Fries, Mycol. vol. i. p. 417.

Found in the vicinity of Downton by DR. DICKSON.

P. saxifraga. Smith, E. B. vol. ii. p. 89. Sowerby, E. B. vol. vi. pl. 407.

Pastures between Salisbury and Stratford. Hedge-banks near Wick.

PICRIS—OX-TONGUE.

P. echioides. (Bristly ox-tongue.) Smith, E. B. vol. iii. p. 239. Sowerby, E. B. vol. xiv. pl. 972.

Hedge-rows, near Farley.

PIMPINELLA.

P. magna. (Great burnet saxifrage.) Smith, E. B. vol. ii. p. 90. Sowerby, E. B. vol. vi. pl. 408.

Hedge-rows, near Downton.

PINGUICULA—BUTTER-WORT.

P. lusitanica. Smith, E. B. vol. i. p. 28. Sowerby, E. B. vol. iii. pl. 145.

Bogs, Alderbury common.

POLYGALA—MILK-WORT.

P. vulgaris. Smith, E. B. vol. iii. p. 258. Sowerby, E. B. vol. ii. p. 76.

In heathy pastures, frequent.

POLYGONUM.

P. amphibium. (Amphibious persicaria.) Smith, E. B. vol. ii. p. 232. Sowerby, E. B. vol. vii. pl. 435.

Sides of rivulets near Downton.

P. Convolvolus. (Climbing back-wheat, or bindweed.) Smith, E. B. vol. ii. p. 240. Sowerby, E. B. vol. xiv. pl. 941.

In corn-fields, entwining itself around wheat and other plants.

P. Hydropifer. (Biting persicaria.) Smith, E. B. vol. ii. p. 235. Sowerby, E. B. vol. xiv. pl. 989.

River-sides, between Downton and Charford Farm.

P. juniperinum. Hooker and Taylor, Musc. Brit. p. 45, tab. 10.

Heath, near Downton. [Dr. Dickson.]

P. Persicaria. (Spotted persicaria.) Smith, E. B. vol. ii. p. 233. Sowerby, E. B. vol. xi. pl. 756.

By the sides of the Avon, not far from Downton; very abundant.

POLYTRICHUM—GOLDEN-HAIR.

P. commune. (Common golden-hair.) Hooker and Taylor, Musc. Brit. p. 46, tab. 10. Sowerby, E. B. vol. xvii. pl. 1179.

Clarendon Wood.

P. nanum. Hooker and Taylor, Musc. Brit. p. 50, tab. 10. Sowerby, E. B. vol. xxiii. pl. 1625.

Moist banks near Downton.

P. undulatum. Hooker and Taylor, Musc. Brit. p. 43, tab. 10. Sowerby, E. B. vol. xvii. pl. 1220.

Common moist hedge-rows, about Alderbury.

POPULUS—POPLAR.

P. alba. (White poplar.) Smith, E. B. vol. iv. p. 243. Sowerby, E. B. vol. xxiii. pl. 1618.

Inner Foss, Old Sarum.

P. tremula. (Aspen.) Smith, E. B. vol. iv. p. 244. Sowerby, E. B. vol. 27, pl. 1909.

Hedges, near Alderbury; also not far from Downton.

POTENTILLA.

P. anserina. (Silver weed.) Smith, E. B. vol. ii. p. 417. Sowerby, vol. xii. pl. 861.

Road-sides in moist places, near Downton.

P. argentea. (Hoary cinquefoil.) Smith, E. B. vol. ii. p. 418. Sowerby, E. B. vol. ii. pl. 89.

Between the windmill and the miller's house on the common, near Whaddon.

P. fragariastrum. Smith, E. B. vol. ii. p. 425. Sowerby, E. B. vol. xxv. pl. 1785.

This is the *fragaria sterilis* of Linnæus.

Dry banks, near Redlynch.

P. reptans. (Creeping cinquefoil.) Smith, E. B. vol. ii. p. 423. Sowerby, E. B. vol. xii. pl. 862.

Sides of rivulets, Downton, and near Laverstock.

POTERIUM—COMMON BURNET.

P. sanguisorba. Smith, E. B. vol. iv. p. 147. Sowerby, E. B. vol. xii. pl. 860.

Upland pastures about Clarendon; and on Salisbury Plain.

PRUNELLA.

P. vulgaris. (Self-heal.) Smith, E. B. vol. iii. p. 114. Sowerby, E. B. vol. xiv. pl. 961.

Hedge-rows, very common.

PYRETHRUM—FEVERFEW.

P. Parthenium. Smith, E. B. vol. iii. p. 451. Sowerby, E. B. vol. xviii. pl. 1231.

Moist lanes, near Downton.

PYRUS.

P. Malus. (Crab-tree.) Smith, E. B. vol. ii. p. 362. Sowerby, E. B. vol. iii. pl. 179.

Hedges, at a place called the *Three Gates*, near Downton.

QUERCUS—OAK.

Q. robur. (Common oak.) Smith, E. B. vol. iv. p. 148. Sowerby, E. B. vol. xix. pl. 1342.

This species is very common, but no where remarkably flourishing (in a wild state), according to my observation, in the vicinity of Salisbury.

RAMALINA.

R. fastigiata.

β. calicaris. Acharii Lich. Univ. p. 604.

About Downton. [Dr. Dickson.]

RANUNCULUS—crowfoot.

R. aquatilis. (Floating crowfoot.) Smith, E. B. vol. iii. p. 54. Sowerby, vol. ii. pl. 101.

Shallow rivulets, near Downton, and in many other places.

R. Flammula. (Lesser spear-wort.) Smith, E. B. vol. iii. p. 45. Sowerby, E. B. vol. vi. pl. 387.

Boggy ground, near Alderbury.

R. hederaceus. (Ivy-leaved crowfoot.) Smith, E. B. vol. iii. p. 54. Sowerby, E. B. vol. xxviii. pl. 2003.

Ditches, near Alderbury, and also near Downton, not uncommon.

R. repens. (Creeping crowfoot.) Smith, E. B. vol. iii. p. 51. Sowerby, E. B. vol. viii. pl. 516.

Alderbury common.

R. sceleratus. (Celery-leaved crowfoot.) Smith, E. B. vol. iii. p. 48. Sowerby, E. B. vol. xix. pl. 681.

Ditch-banks, Fisherton.

RESEDA—rocket.

R. lutea. (Wild mignionette.) Smith, E. B. vol. ii. p. 348. Sowerby, E. B. vol. v. pl. 321.

On hillocks of chalk around the lime-kilns near Mr. Wyndham's grounds, Salisbury, abundant. Also near Wick.

R. luteola. (Dyer's weed.) Smith, E. B. vol. ii. p. 347. Sowerby, E. B. vol. v. pl. 320.

Road-sides, near Laverstock. This plant grows to an enormous neight, and very abundantly, near a chalk-pit at Wick, not far from the "Giant's Chair."

RHINANTHUS—rattle.

R. Crista-galli. Smith, E. B. vol. iii. p. 120. Sowerby, E. B. vol. x. pl. 657.

Common in meadows.

RHIZOMORPHA.

R. Hypoxylon.

Moist banks, Alderbury.

RHYNCHOSPORA.

R. alba. (White bog-rush.) Smith, E. B. vol. i. p. 52. Sowerby, E. B. vol. xiv. pl. 985.

The *schœnus albus,* of Linnæus.

In bogs, Alderbury Common.

RIBES—CURRANT.

R. Grossularia. (Gooseberry.) Smith, E. B. vol. i. p. 333. Sowerby, E. B. vol. xviii. pl. 1292.

I have seen this species by the road-sides, between Downton and Wick, but the situation is suspicious, and indeed a *quære* remains attached to it, in the British Flora, as to its claim to be considered indigenous in our island.

R. rubrum. (Red Currant.) Smith, E. B. vol. i. p. 330. Sowerby, E. B. vol. xviii. pl. 1289.

Rather abundant, in a lane leading from the Netherhampton Road towards the race-plain.

ROSA—ROSE.

R. arvensis. (Field-rose.) Smith, E. B. vol. ii. p. 396. Sowerby, E. B. vol. iii. pl. 188.

Borders of fields, about Old Sarum.

R. canina. (Dog-rose.) Smith, E. B. vol. ii. p. 394 Sowerby, E. B. vol. xiv. pl. 992.

Very frequent in hedges.

RUBUS.

R. cœsius. (Dew-berry.) Smith, E. B. vol. ii. p. 409. Sowerby, E. B. vol. xii. pl. 826.

Woody spots, near Winterslow. I have usually found it flowering in August.

RUMEX.

R. acetosa. (Sorrell.) Smith, E. B. vol. ii. p. 196. Sowerby, E. B. vol. ii. pl. 127.

Pastures, about Salisbury.

R. hydrolapathum. (Water-dock.) Smith, E. B. vol. ii. p. 196. Sowerby, E. B. vol. xxx. pl. 2104.

In rivulets about Downton, very frequent.

R. palustris. Smith, E. B. vol. ii. p. 194. Sowerby, E. B. vol. xxvii. pl. 1392.

Meadows, rather abundant.

SAGITTARIA.

S. sagittifolia. (Arrow-leaf.) Smith, E. B. vol. iv. p. 145. Sowerby, E. B. vol. ii. pl. 84.

This plant is very common in rivulets communicating with the Avon.

SALIX.

S. caprea. (Round-leaved sallow.) Smith, E. B. vol. iv. p. 225, Sowerby, E. B. vol. xxi. pl. 1488.

This species is more commonly called the *broad-leaved willow,* and in the fen countries is the one, the bark of which is used, and with success, in the cure of agues. It may be found by the river-sides about Britford.

S. Lambertiana. Smith, E. B. vol. iv. p. 189. Sowerby, E. B. vol. xix. pl. 1359.

On the banks of the Willey for the course of sixteen miles. [A. B. Lambert, Esq.—as quoted in Smith's E. B.]

S. repens. (Creeping willow.) Smith, E. B. vol. iv. p. 209. Sowerby, E. B. vol. iii. pl. 183.

Bogs, Alderbury Common.

S. rubra. (Green-leaved osier.) Smith, E. B. vol. iv. p. 191. Sowerby, E. B. vol. xvi. pl. 1145.

In osier grounds, near Salisbury. (*Gough's Camden.*)

" By the side of the river near Salisbury." (*Withering.*)

The first of these habitats is certainly, and the latter most probably, on the authority of Ray; but it is given so vaguely as to afford no guidance to the spot. There are some osier-grounds between Salisbury and Bemerton, where this species is likely to be found.

" It is much sought after by basket-makers." (*Withering.*)

SALVIA—SAGE.

S. verbenaca. (Wild clary.) Smith, E. B. vol. i. p. 35. Sowerby, E. B. vol. iii. pl. 154.

On Milford Hill, abundant.

SAMBUCUS—ELDER.

S. Ebulus. (Dwarf elder.) Smith, E. B. vol. ii. p. 108. Sowerby, E. B. vol. vii. pl. 475.

In dry trenches, near Combe, and also near Redlynch.

S. nigra. (Common elder.) Smith, E. B. vol. ii. p. 109. Sowerby, E. B. vol. vii. pl. 476.

Common in hedges.

SANICULA—SANICLE.

S. europæa. Smith, E. B. vol. ii. p. 36. Sowerby, E. B. vol. ii. pl. 98.

Groves of Clarendon.

SAPONARIA—SOAP-WORT.

S. officinalis. Smith, E. B. vol. ii. p. 284. Sowerby, E. B. vol. xv. pl. 1060.

Ditch-banks, near West Harnham.

SATYRIUM.

S. viride.

In a pasture near Eyre's Gutter-gate, not far from Trafalgar Park.

SAXIFRAGA—SAXIFRAGE.

S. granulata. Smith, E. B. vol. ii. p. 269. Sowerby, E. B. vol. vii. pl. 500.

Meadows, near Alderbury.

S. tridactylites. Smith, E. B. vol. ii. p. 271. Sowerby, E. B. vol. vii. pl. 501.

On walls, common.

SCABIOSA—SCABIOUS.

S. columbaria. (Small scabious.) Smith, E. B. vol. i. p. 195. Sowerby, E. B. vol. xix. pl. 1311.

Hill, near the lime-kiln, Milford.

S. succisa. (Devil's bit.) Smith, E. B. vol. i. p. 194. Sowerby, E. B. vol. xiii. pl. 878.

In a meadow, to the right of the lane leading from Alderbury to Trafalgar Park, abundantly.

SCANDIX.

S. Pecten Veneris. (Venus's comb.) Smith, E. B. vol. ii. p. 17. Sowerby, E. B. vol. xx. pl. 1397.

Corn-fields, not unfrequent.

SCIRPUS—CLUB-RUSH.

S. lacustris. Smith, E. B. vol. i. p. 56. Sowerby, E. B. vol. x. pl. 666.

Deeper rivulets, communicating with the Avon, near Stratford.

SCLERANTHUS—KNAWELL.

S. annuus. (Annual knawell.) Smith, E. B. vol. ii. p. 283. Sowerby, E. B. vol. v. pl. 351.

Corn-fields, in sandy spots, near Clerebury Camp.

SCOLOPENDRIUM—HART'S-TONGUE.

S. vulgare. Smith, E. B. vol. iv. p. 314. Sowerby, E. B. vol. xvi. pl. 1150.

Asplenium scolopendrium of Linnæus.

Shady banks, near Milford; and, also, between Downton and Redlynch.

SCROPHULARIA.

S. aquatica. (Water figwort.) Smith, E. B. vol. iii. p. 138. Sowerby, E. B. vol. xii. pl. 854.

River-sides, near the road from Salisbury to Alderbury.

SCUTELLARIA—SKULL-CAP.

S. galericulata. Smith, E. B. vol. iii. p. 112. Sowerby, E. B. vol. viii. pl. 523.

River-sides, near Britford, and in a meadow, near Downton.

S. minor. Smith, E. B. vol. iii. p. 112. Sowerby, E. B. vol. viii. pl. 524.

Moist ground, on Alderbury Common.

SEDUM.

S. sexangulare. Smith, E. B. vol. ii. p. 318. Sowerby, E. B. vol. xxviii. pl. 1946.

On the ruins of the walls of Old Sarum. Here it was found also by Mr. Dawson Turner and Mr. Sowerby.

SEMPERVIVUM.

S. tectorum. (House-leek.) Smith, E. B. vol. ii. p. 350. Sowerby, E. B. vol. xix. pl. 1320.

On walls, not unfrequent.

SENECIO—RAGWORT.

S. aquaticus. (Marsh ragwort.) Smith, E. B. vol. iii. p. 434. Sowerby, E. B. vol. xvi. pl. 1131.

Meadows, near the turnpike-gate called Peter's Finger, on the road from Salisbury to Alderbury.

S. viscosus. (Stinking groundsel.) Smith, E. B. vol. iii. p. 429. Sowerby, E. B. vol. i. pl. 32.

On chalky hillocks, near Alderbury.

SERRATULA—SAW-WORT.

S. tinctoria. (Dyer's saw-wort.) Smith, E. B. vol. iii. p. 382. Sowerby, E. B. vol. i. pl. 38.

On the hill, in sandy spots, near Dinton.

SHERARDIA.

S. arvensis. Smith, E. B. vol. i. p. 196. Sowerby, E. B. vol. xiii. pl. 891.

In upland chalky corn-fields, west of Downton.

SILENE—CATCH-FLY.

S. conica. Smith, E. B. vol. ii. p. 294. Sowerby, E. B. vol. xiii. pl. 922.

In a lane between Downton and Charlton.

I have put a quære against this species, because the only specimens which I could find were in an imperfect and mutilated state, and I was not able to name them with positive *certainty*.

SINAPIS—MUSTARD.

S. alba. (White mustard.) Smith, E. B. vol. iii. p. 222 Sowerby, E. B. vol. xxiv. pl. 1677.

On mud-walls and among rubbish, frequent.

S. nigra. (Black mustard.) Smith, E. B. vol. iii. p. 222. Sowerby, E. B. vol. xiv. pl. 969.

Road-sides, about Laverstock.

SISON—HONEWORT.

S. Amomum. Smith, E. B. vol. ii. p. 60. Sowerby, E. B. vol. xiv. pl. 954.

Under hedges, in moist places, about Downton.

SISYMBRIUM.

S. Irio. Smith, E. B. vol. iii. p. 197. Sowerby, E. B. vol. xxiii. pl. 1631.

Sides of rivulets, near Upper Charlton Farm.

SIUM.

S. angustifolium. (Narrow-leafed water-parsnip.) Smith, E. B. vol. ii. p. 56. Sowerby, E. B. vol. ii. p. 139.

Rivulets, near Charlton.

SOLANUM.

S. Dulcamara. (Bitter-sweet.) Smith, E. B. vol. i. p. 317. Sowerby, E. B. vol. viii. pl. 365.

Hedges, near Downton, also, near Dinton.

S. nigrum. Smith, E. B. vol. i. p. 319. Sowerby, E. B. vol. viii. pl. 566.

Banks, under hedges, near Alderbury, and about Redlynch.

SOLIDAGO.

S. Virga aurea. (Golden rod.) Smith, E. B. vol. iii. p. 438. Sowerby, E. B. vol. v. pl. 301.

This showy plant is not uncommon in woody lanes about Alderbury and Downton.

SPARTIUM—BROOM.

S. scoparium. Smith, E. B. vol. iii. p. 261. Sowerby, E. B. vol. xix. pl. 1339.

In upland sandy spots, near Dinton.

SPERGULA—SPURREY.

S. nodosa. (Knotted spurrey.) Smith, E. B. vol. ii. p. 338. Sowerby, E. B. vol. x. pl. 694.

In moistish spots, on Alderbury Common.

SPHÆRIA.

S. sanguinea. Fries, Syst. Mycol. vol. ii. p. 455. Sowerby, Fungi, vol. iii. pl. 254.

On rotten fragments of ash, and cherry trees.

S. tremelloides. Fries, Syst. Mycol. vol. ii. p. 335.

This rich purple-coloured species occurs not unfrequently on damp, decaying wood.

SPHAGNUM.

S. acutifolium. Hooker and Taylor, Musc. Brit. p. 14, tab. 4.

Near Downton. [DR. DICKSON.]

SPIRÆA.

S. *Filipendula.* (Dropwort.) Smith, E. B. vol. ii. p. 368. Sowerby, E. B. vol. iv. pl. 284.

Downs, near Stonehenge.

S. Ulmaria. (Meadow-sweet.) Smith, E. B. vol. ii. p. 368. Sowerby, E. B. vol xiv. pl. 960.

In meadows, common.

SPLACHNUM.

S. ampullaceum. Hooker and Taylor, Musc. Brit. p. 39, tab. 9. Sowerby, E. B. vol. ii. pl. 144.

Bogs, Alderbury Common.

STACHYS.

S. arvensis. Smith, E. B. vol. iii. p. 100. Sowerby, E. B. vol. xvii. pl. 1154.

Corn-fields, near Stratford.

S. palustris. Smith, E. B. vol. iii. p. 99. Sowerby, E. B. vol. xxiv. pl. 1675.

Stratford Marsh.

S. sylvatica. Smith, E. B. vol. iii. p. 98. Sowerby, E. B. vol. vi. pl. 416.

Hedge-rows, about Downton, and elsewhere in this district.

STELLARIA—STITCHWORT.

S. graminea. Smith, E. B. vol. ii. p. 302. Sowerby, E. B. vol. xii. pl. 803.

Woods, near Winterslow; hedge-rows, near Redlynch.

S. holostea. Smith, E. B. vol. ii. p. 301. Sowerby, E. B. vol. viii. pl. 511.

Clarendon Wood.

S. uliginosa. Smith, E. B. vol. ii. p. 302. Sowerby, E. B. vol. xv. pl. 1074.

Moist parts of West Dean Woods.

STICTA.

S. pulmonacea. Acharii Lich. Univ. p. 449. Sowerby, E. B. vol. viii. pl. 572.

Trunks of lime-trees, in Clarendon Wood.

This is the *lichen pulmonarius* of Linn.

TAMUS.

T. communis. (Black Bryony.) Smith, E. B. vol. ii. p. 241. Sowerby, E. B. vol. ii. pl. 91.

Hedges, about Alderbury and Downton.

TAXUS—YEW.

T. baccata. Smith, E. B. vol. iv. p. 253. Sowerby, E. B. vol. xi. pl. 746.

Not uncommon on the higher chalky grounds in the vicinity of Salisbury.

Mr. Aubrey remarks (in his MS. work on Wiltshire), that in his time, "there were between Merton (vulgo Martin) and Downeton, on the hills, woods of Yewgh Trees." Certainly, even now, this tree is more frequent and flourishing about the last-mentioned place than in any other part of the county.

TEUCRIUM.

T. Scorodonia. (Wood-sage.) Smith, E. B. vol. iii. p. 68. Sowerby, E. B. vol. xxii. pl. 1543.

Road-sides, about Farley.

THALICTRUM—MEADOW-RUE.

T. flavum. Smith, E. B. vol. iii. p. 42. Sowerby, E. B. vol. vi. pl. 367.

Ditch-banks, near Fisherton-Anger, and near Downton, abundantly.

THELEPHORA.

T. lactea. (Fibrillaria stellata.) Fries, Mycol. vol. i. p. 452. Sowerby, Fungi, pl. 387, f. 1.

Collected near Downton by DR. DICKSON.

THESIUM.

T. linophyllum. Smith, E. B. vol. i. p. 337. Sowerby, E. B. vol. iv. pl. 247.

Found on the high chalky grounds above Odstock. [Mr. Roberts, A.L.S.]

Salisbury Plain. [Gough's Camden.]

THYMUS—thyme.

T. Acinos. (Basil thyme.) Smith, E. B. vol. iii. p. 109. Sowerby, E. B. vol. vi. p. 411.

On elevated chalky spots, especially by road-sides. Not uncommon about Pitton and Downton.

TORDYLIUM—hart-wort.

T. maximum. Smith, E. B. vol. ii. p. 105. Sowerby, E. B. vol. xvii. pl. 1173.

This plant is considered by English botanists rare. I have found it on the borders of a corn-field, near a hedge-row, between Downton and Charlton, where it grows, (though sparingly), to a great height.

TORTULA.

T. subulata. Hooker and Taylor, Musc. Brit. p. 57. tab. 12. Sowerby, E. B. vol. xvi. pl. 1101.

Dry banks, near Alderbury.

T. unguiculata. Hooker and Taylor, Musc. Brit. p. 57, tab. 12. Sowerby, E. B. vol. xxxiii. pl. 2316.

Banks, near Downton. [Dr. Dickson.]

TRAGOPOGON—goats-beard.

T. pratensis. Smith, E. B. vol. iii. p. 337. Sowerby, E. B. vol. vii. pl 434.

Sides of the path leading from Swayne's Close towards Old Sarum.

TREMELLA.

T. albida. Fries, Mycol. vol. ii. p. 215. Sowerby, E. B. vol. xxx. pl. 2117.

T. mesenterica. Fries, Mycol. vol. ii. p. 214. Sowerby, E. B. vol. x. pl. 709.

Both of the above species have been found in the vicinity of Downton, by Dr. Dickson.

TRIFOLIUM—TREFOIL.

T. officinale. (Common Melilot.) Smith, E. B. vol. iii. p. 297. Sowerby, E. B. vol. xix. pl. 1340.

Sides of a lane leading from Alderbury to Trafalgar Park.

T. procumbens. (Procumbent trefoil.) Smith, E. B. vol. iii. p. 309. Sowerby, E. B. vol. xiv. pl. 945.

Borders of corn-fields, in gravelly spots, but not very common.

T. subterraneum. (Subterraneous trefoil.) Smith, E. B. vol. iii. p. 300. Sowerby, E. B. vol. xv. pl. 1048.

This little plant is abundant on the higher and drier parts of Alderbury Common.

TRIGLOCHIN.

T. palustre. (Marsh Arrow Grass.) Sowerby, E. B. vol. vi. pl. 366.

Bogs, Alderbury Common.

TUBER.

T. cibarium. (Trufle.) Sowerby, Fungi, vol. iii. pl. 39.

In the ground, in beech woods, about Winterslow, where it is dug up by dogs. [P. B. Brodie, Esq.]

TUSSILAGO.

T. Petasites. (Colt's-foot.) Smith, E. B. vol. iii. p. 425. Sowerby, E. B. vol. vi. pl. 431.

Specimens of this plant were brought to me from the sides of the river at Bemerton, by the late Mr. Thomas Lake.

TYPHA—REED-MACE.

T. latifolia. (Broad-leaved reed-mace.) Smith, E. B. vol. iv. p. 71. Sowerby, E. B. vol. xxi. pl. 1455.

Stratford Marsh.

USNEA.

U. barbata. var. γ. articulata. Acharii, Lich. Univ. p. 625.

Branches of decaying oak trees, at Winterslow.

UTRICULARIA—BLADDER-WORT.

U. minor. (Lesser bladder-wort.) Smith, E. B. vol. i. p. 31. Sowerby, E. B. vol. iv. p. 254.

This scarce little plant may be found in rivulets near the Milk-maid's Grove, close to Salisbury.

VALERIANA—VALERIAN.

V. dioica. (Small Marsh Valerian.) Smith, E. B. vol. i. p. 43. Sowerby, E. B. vol. ix. pl. 628.

Meadows, about Downton.

V. officinalis. (Medicinal valerian.) Smith, E. B. vol. i. p. 43. Sowerby, E. B. vol. x. p. 698.

In boggy pastures about Alderbury and Downton.

VARIOLARIA.

V. communis. β. *faginea.* Acharii, Lich. Univ. p. 323. Sowerby, E. B. vol. xxiv. pl. 1713.

Lichen fagineus. Linn.
On ash trees, not uncommon.

VELLA.

V. annua. (Cress-rocket.) Sowerby, E. B. vol. xxi. pl. 1442.

" On Salisbury Plain, not far from Stonehenge." [RAY, on the authority of Lawson.]

This is the only spot in England, I believe, where *V. annua* is said to have been found. I have sought for it (as many botanists have done) in vain.

VERBASCUM—MULLEIN.

V. Blattaria. (Moth Mullein.) Smith, E. B. vol. i. p. 312. Sowerby, E. B. vol. vi. pl. 393.

Lanes, between Downton and Charlton.

V. nigrum. Smith, E. B. vol. i. p. 311. Sowerby, E. B. vol. i. pl. 59.

Road-sides, between Salisbury and Milford.

VERBENA—VERVAIN.

V. officinalis. Smith, E. B. vol. iii. p. 71. Sowerby, E. B. vol. xi. pl. 767.

By road-sides, very common.

VERONICA.

V. Anagallis. (Water Speedwell.) Smith, E. B. vol. i. p. 21. Sowerby, E. B. vol. xi. pl. 781.

Ditches, near Milford.

V. arvensis. (Wall Speedwell.) Smith, E. B. vol. i. p. 24. Sowerby, E. B. vol. xi. pl. 734.

On walls, near Salisbury.

V. Beccabunga. (Brook-lime.) Smith, E. B. vol. i. p. 20. Sowerby, E. B. vol. x. pl. 655.

In ditches, very frequent.

V. chamadrys. (Germander Speedwell.) Smith, E. B. vol. i. p. 23. Sowerby, E. B. vol. ix. p. 623.

Common by the sides of roads, in elevated situations, around Salisbury.

V. hederifolia. (Ivy-leaved Speedwell.) Smith, E. B. vol. i. p. 25. Sowerby, E. B. vol. xi. pl. 784.

Sides of corn-fields, among gravel.

V. officinalis. (Common Speedwell.) Smith, E. B. vol. i. p. 22. Sowerby, E. B. vol. xi. pl. 765.

Alderbury Common.

V. scutellata. (Marsh Speedwell.) Smith, E. B. vol. i. p. 21. Sowerby, E. B. vol. xi. pl. 782.

Bogs, near West Dean, and on Alderbury Common.

V. serpyllifolia. (Smooth Speedwell.) Smith, E. B. vol. i. p. 20. Sowerby, E. B. vol. xv. pl. 1075.

Alderbury Common.

VIBURNUM.

V. Lantana. (Mealy tree.) Smith, E. B. vol. ii. p. 107. Sowerby, E. B. vol. v. pl. 331.

This shrub, partial to calcareous soils, abounds in the hedges about Salisbury.

V. Opulus. (Guelder rose.) Smith, E. B. vol. ii. p. 107. Sowerby, E. B. vol. v. pl. 332.

Hedges, in damp situations, near West Harnham and Laverstock.

VICIA—VETCH.

V. Cracca. (Tufted Vetch.) Smith, E. B. vol. iii. p. 280. Sowerby, E. B. vol. xvii. pl. 1168.

Hedge-rows, by the sides of rivulets, near Harnham.

V. lutea. (Rough-podded Yellow Vetch.) Smith, E. B. vol. iii. p. 284. Sowerby, E. B. vol. vii. pl. 481.

This is usually considered a maritime species; but, unless I was greatly mislead by the mutilated state of the few plants which I saw, it grows in a stony spot, on the hill, near an antient earth-work called the "Giant's Chair," at a short distance from the village of Wick. Mr. Dawson Turner also found it in an inland situation, viz., near Glastonbury Torr.

V. sativa. (Common Vetch.) Smith, E. B. vol. iii. p. 281. Sowerby, E. B. vol. v. pl. 334.

I have seen this species chiefly in wooded spots, about Salisbury, but not in profusion.

V. sepium. (Bush Vetch.) Smith, E. B. vol. iii. p. 286. Sowerby, E. B. vol. xxii. pl. 1515.

In hedges.

VINCA—PERIWINKLE.

V. major. (Greater periwinkle.) Smith, E. B. vol. i. p. 339. Sowerby, E. B. vol. viii. pl. 514.

Under hedges, near Milford.

V. minor. (Lesser periwinkle.) Smith, E. B. vol. i. p. 338. Sowerby, E. B. vol. xiii. pl. 917.

This species (much less frequent than its congener,) I have found on shady banks, near Milford Bridge.

VIOLA—VIOLET.

V. canina. (Common dog-violet.) Smith, E. B. vol. i. p. 303. Sowerby, E. B. vol. ix. pl. 620.

Frequent on banks, under hedge-rows.

V. odorata. (Sweet-smelling violet.) Smith, E. B. vol. i. p. 301. Sowerby, E. B. vol. ix. pl. 619.

I have found this species, here and there, in the vicinity of Salisbury, under hedge-rows; but I have not noted the particular spots.

A *white* variety, both of this and of the preceding species, I have sometimes seen in chalky situations.

V. tricolor. (Three-coloured violet, Pansy, or Heartsease.) Smith, E. B. vol. i. p. 305. Sowerby, E. B. vol. xviii. pl. 1287.

Corn-fields, in higher situations, common.

VISCUM—MISSELTOE.

V. album. (White misseltoe.) Smith, E. B. vol. iv. p. 236. Sowerby, E. B. vol. xxi. pl. 1470.

Though Wiltshire may be considered a *Druidical* county, I have not seen the misseltoe growing in it in any abundance.

In gardens it is found on apple trees. About Whiteparish, I have gathered it from the white-thorn.

WEISSIA.

W. controversa. Hooker and Taylor, Musc. Brit. p. 84, tab. 15. Road-sides, near Downton.

AVES.

ALCEDO—KING-FISHER.

A. Ispida. Linn. Syst. Nat. à Gm. Penn. Br. Zool. ed. 4. vol. i. p. 246. pl. 38. No. 89.

This interesting and beautiful bird is often seen on the banks of the rivulets about Salisbury.

ARDEA.

A. Ciconia. (White Stork.) Linn. Syst. Nat. à Gm.

This is a rare bird in England, but we are informed by Colonel Montagu (Ornithological Dictionary, vol. ii.) that one of them was killed at Salisbury, in February, 1790.

A. major. (Common Heron.) Linn. Syst. Nat. à Gm. Penn. Br. Zool. ed. 4. vol. ii. p. 421. pl. 61. No. 173.

Seen in the meadows near Salisbury; shot there in January, 1795. Also near Downton.

A. stellaris. (Bittern.) Linn. Syst. Nat. à Gm. Penn. Br. Zool. ed. 4. vol. i. p. 424.

Seen in considerable numbers among rushes, near Britford, in January, 1795. There was a very hard frost at this time.

CAPRIMULGUS—GOAT-SUCKER.

C. europæus. (Nocturnal goat-sucker.) Linn. Syst. Nat. à Gm. Penn. Br. Zool. ed. 4. vol. i. p. 416. pl. 59. No. 172.

Brought to me from Coombe, in September, 1796.

The absurdity of the name of this bird ought to be here noted; but the error of supposing that it sucks goats is as old as the days of Aristotle.

CHARADRIUS.

C. Morinellus. (Dotterel.) Linn. Syst. Nat. à Gm. Penn. Br. Zool. ed. 4. vol. ii. p. 477. pl. 73. No. 210.

Seen on Uphaven Downs. About Amesbury. [AUBREY.]

C. œdicnemus. (Thick-kneed plover.) Linn. Syst. Nat. à Gm.

In Wiltshire very generally called the "Thick-kneed *Bustard*," though the bustard is of a very different genus. One of these birds* was brought to me in August, 1796, from a heath near Gumbleton (about six miles from Salisbury), where it was taken up in a very young state. I fed it with raw flesh about a fortnight; at the end of that time, being set at liberty in my father's garden, at Salisbury, it lived chiefly on worms and insects, but was found dead on the 7th of December that year, having been killed apparently by a most severe frost.

COLUMBA—PIGEON.

C. Turtur. (Turtle-dove.) Linn. Syst. Nat. à Gm. Penn. Br. Zool. ed. 4. vol. i. p. 297.

Aubrey says that, in his time, the turtle-dove was seen "about Ivy Church, and so to Whiteparish in the chalkie waies, but they are not very common." The bird is now and then seen there still, as I have understood from residents in the neighbourhood.

COLYMBUS—GREBE.

C. auritus. γ. (Didapper.) Linn. Syst. Nat. à Gm. Penn. Br. Zool. ed. 4. vol. ii. p. 501.

River Avon, about Stratford.

C. cristatus. (Crested grebe.) Linn. Syst. Nat. à Gm. Penn. Br. Zool. ed. 4. vol. ii. p. 497.

This bird has been shot in the immediate vicinity of Salisbury, but I have seen only one specimen, and this was in a very hard winter.

FALCO.

F. Milvus. (Kite.) Linn. Syst. à Gm. Penn. Br. Zool. ed. 4. vol. i. p. 185.

Seen sometimes about Downton.

FULICA.

F. Chloropus. (Water Hen Coot.) Linn. Syst. Nat. à Gm.

Stratford Marsh.

* On comparing this individual with the description of the species in Pennant's Zoology, I perceived these differences, viz: there were *three* black quill feathers, and the tips of the *three* outmost tail-feathers were of the same colour. The eyes of this bird are remarkably *full*, and have the appearance of being immoveable in their sockets.

I have not inserted a reference to Pennant under the head of *C. œdicnemus*, that author being very incorrect in the nomenclature, as well as in the characters which he has assigned to this species.

AVES.

HIRUNDO.

H. riparia. (Land Swallow.) Linn. Syst. Nat. à Gm. Penn. Br. Zool. ed. 4. vol. i. p. 402.

This bird builds in the sand which overlays some of the chalk-pits about Downton, in which neighbourhood the poles are very numerous, though *H. riparia* is a much less common species than *H. urbica* (the martin).

LARUS.

L. ridibundus. (Pewit.) Linn. Syst. Nat. à Gm. Penn. Br. Zool. ed. 4. vol. ii. p. 541.

This bird has been seen on the downs between Salisbury and Newton Tony. An old sportsman, the late Mr. Dew, informed me that he saw several near the last-mentioned place in a hard winter.

MOTACILLA.

M. Baarula. (Grey Wagtail.) Linn. Syst. à Gm. Penn. Br. Zool. ed. 4. vol. i. p. 363.

This bird is not uncommon, by the sides of rivulets, in the vicinity of Salisbury. I have seen it most commonly in meadows, near St. Thomas's bridge.

RULLUS—RAIL.

R. Crex. (Land-rail.) Lynn. Syst. Nat. à Gm. Penn. Br. Zool. ed. 4. vol. ii. p. 487. pl. 751. No. 216.

This bird was very common in corn-fields about Salisbury, in my younger days; but I have understood that they do not abound there now.

SCOLOPAX.

S. arquata. Linn. Syst. Nat. à Gm. Penn. Br. Zool. ed. 4. vol. ii. p. 429. pl. 63.

My friend the Rev. Edward Bowen informed me that he once saw some of these birds on the plain not far from Salisbury.

STRIX—OWL.

S. Aluco. β. stridula. (Screech owl.) Linn. Syst. Nat. à Gm. Penn. Br. Zool. ed. 4. vol. i. p. 208.

Not uncommon about Salisbury, where it is sometimes heard from the chimney-tops. The vulgar are still much alive to its cries, considering them as portending death to some individual in the family.

TETRAO.

T. Coturnix. (Quail.) Linn. Syst. Nat. à Gm. Penn. Br. Zool. ed. 4. vol. i. 276.

These birds were very numerous in pastures about Bishop's Down, when I was young. I have often caught them with a (green silk) net and a call-pipe, kept them in cages during the winter, and turned them out again in good condition in the spring; but I believe that they are now rare, at least in this district.

TRINGA.

T. Vanellas. (Lapwing.) Linn. Syst. Nat. à Gm. Penn. Br. Zool. ed. 4. vol. ii. p. 453.

Not unfrequent on the Downs about Salisbury.

TURDUS.

T. Merula. (Blackbird.) Linn. Syst. Nat. à Gm. Penn. Br. Zool. ed. 4. vol. i. p. 308. pl. 47. No. 109.

This bird haunts plantations about Downton and Redlynch in great numbers, and is much detested by gardeners.

UPUPA—HOOPOE.

U. Epops. Linn. Syst. Nat. à Gm. Penn. Br. Zool. ed. 4. vol. i. p. 257. pl. 39. No. 90.

This bird is ascertained to have been shot near Downton. [GEORGE MATCHAM, ESQ.]

YUNKS—WRYNECK.

Y. Torquilla. Linn. Syst. Nat. à Gm. Penn. Br. Zool. ed. 4. vol. i. p. 237. pl. 36. No. 83.

An individual of this species, shot near Winterbourn Earl's by Mr. Cust, was brought to me in August, 1796. I believe that *Y. Torquilla* is seen in Wiltshire *very rarely*.

AMPHIBIA.

ANGUIS.

A. fragilis. (Slow Worm, or Blind Worm.) Linn. Nat. à Gm. Penn. Br. Zool. ed. 4. vol. iii. p. 36. pl. 4. No. 15.

This creature is very aptly denominated "*fragilis;*" I can myself bear testimony to its brittleness, having seen it broken short into fragments quite flat at the extremities by slight blows of a stick. It is sometimes confounded with the Adder *(Coluber Berus)*, and hence been considered venomous, though I believe it to be perfectly innocuous. Its motions are remarkably slow, and from which circumstance it has got the names of *Slow-worm* and *Blind-worm*. The former is its more common appellation in Wiltshire. I have seen it only on very dry, chalky hillocks, on the sides of the road near the foot of Alderbury Hill.

COLUBER—SERPENT.

C. Berus. (Adder.) Linn. Syst. Nat. à Gm. Penn. Br. Zool. ed. 4. vol. iii. p. 26.

This serpent is sometimes seen of an almost uniformly brownish colour, and, when of this appearance, and also very young, it is confounded by the vulgar with the Blind Worm *(Anguis fragilis)* above described. It is not unfrequent on dry, chalky hillocks, about Salisbury.

C. Natrix. (Snake.) Linn. Syst. Nat. à Gm. Penn. Br. Zool. ed. 4. vol. iii. p. 33.

In woods and hedges. Pennant has given pictures of both these species of Coluber, but I do not consider them sufficiently accurate for reference.

LACERTA—LIZARD.

L. agilis. Linn. Syst. Nat. à Gm. Penn. Br. Zool. ed. 4. vol. iii. p. 21. pl. 2. No. 9.

The only place where I have noticed this little animal in Wiltshire is about Fisherton-Anger, on dry banks.

MAMMALIA.

ERINACEUS—HEDGEHOG.

D. Europæus. (Common hedge-hog.) Linn. Syst. Nat. à Gm. Penn. Brit. Zool. ed. 4. vol. i. p. 133.

Hedge-rows between Milford and Clarendon Park.

MUSTELA.

M. Lutra. (Otter.) Linn. Syst. Nat. à Gm. Penn. Br. Zool. ed. 4. vol. i. p. 92, pl. 8, n. 19.

River Avon, near New Farm, Downton.

M. vulgaris. (Weasel.) Penn. Brit. Zool. ed. 4. vol. i. p. 95. pl. 7. No. 17.

About Downton, in barns.

SCIURUS—SQUIRREL.

S. vulgaris. Linn. Syst. Nat. à Gm. Penn. Br. Zool. ed. 4. vol. i. p. 107.

Plantations at Redlynch.

SOREX—SHREW.

S. Araneus. Linn. Syst. Nat. à Gm. Penn. Br. Zool. ed. 4. vol. i. p. 125.

This animal is called in Wiltshire the "*Over-runner.*" I have seen it lying dead on dry banks about Alderbury, but never living.

TALPA—MOLE.

T. Europæa. (Common Mole.) Linn. Syst. Nat. à Gm. Penn. Br. Zool. ed. 4. vol. i. p. 128.

Alderbury common. Dr. Maton's grounds, at Redlynch.

PISCES.

PERCA—PERCH.

P. fluviatilis. (River-perch.) Linn. Syst. Nat. à. Gm. Penn. Br. Zool. ed. 4. vol. iii. p. 254. No. 124. pl. 48.

This fish abounds in the Nadder, about Bemerton. I have caught them of a very considerable size there.

SALMO.

S. Fario. (Trout.) *Linn*. Syst. Nat. à Gm. Penn. Br. Zool. ed. 4. vol. iii. p. 297, pl. 59, No. 146.

The trout abounds and attains considerable size, in the streams about Salisbury.

S. Salar. (Salmon.) Linn. Syst. Nat. à Gm. Penn. Br. Zool. ed. 4. vol. iii. p. 284. pl. 58.

Aubrey says that the salmon has been caught (though rarely) about Harnham Bridge. Upon his authority, therefore, I include this species among the Wiltshire fish; but I know no person now living who has ascertained its having ascended the Avon so far as Salisbury.

S. thymallus. (Umber, or Grayling.) Linn. Syst. Nat. à Gm. Penn. Br. Zool. ed. 4. vol. iii. p. 311. pl. 61. No. 150.

This species is still to be found (in the Avon) at Downton, where Walton speaks of its being caught in his time. It is a fish of delicate flavour, and is often brought by the "drowners" (as they are called) of that neighbourhood to my country house.

Aubrey, in his MS., says, that, in his days, the Umber was caught in the Nadder, between Wilton and Salisbury, and that it was sent from the latter place to the metropolis. "This kind of fish (he remarks,) is found in no other river in England, except the Humber, in Yorkshire." From that river, therefore, I conclude that it takes its name of *Umber*.

My late friend Dr. Babington caught it near Ringwood, but this place is on the banks of the Avon.

VERMES.

CARDIUM—COCKLE.

C. corneum. Linn. à Gm. Wood, Ind. Test. pl. 5. fig. 90.

Rivulets, near St. Thomas's Bridge, and in other parts of the vicinity of Salisbury, frequent.

HELIX—SNAIL.

H. arbustorum. Linn. à Gm. Wood, Ind. Test. pl. 34. fig. 88. In hedges, but not very common.

H. auricularia. Linn. à Gm. Wood, Ind. Test. pl. 35. fig. 180. In rivulets and ditches.

H. hispida. (Bristled Snail.) Linn. à Gm. Wood, Ind. Test. pl. 33, fig. 64.

This species is mentioned as being found "at Salisbury," by Mr. Jeffreys, in his "*Synopsis of Testaceous Pneumondranchous Mollusca of Great Britain,*" (Linn. Tr. vol. xvi. p. 507) where it bears the name of *H. globularis.*

H. Lapicida. Linn. à Gm. Wood, Ind. Test. pl. 32. fig. 3. On dry chalky eminences about Alderbury.

H. Planorbis. Linn. à Gm. Linn. Tr. vol. viii. p. 188. pl. 5. fig. 13.

In rivulets, not uncommon; but it is apt to be confounded with *H. planata* and *complanata.*

H. radiata. (Radiated Snail-shell.) Wood, Ind. Test. pl. 32. fig. 11.

Found in various parts of the vicinity of Salisbury, on dry banks; commonly on moss.

H. rufescens. (Reddish Snail.) Linn. Tr. vol. viii. p. 196 Wood, Ind. Test. pl. 33 fig. 20.

Under Hedges, not uncommon. In the neighbourhood of Salisbury. [MR. JEFFREYS, in Linn. Trans.]

H. succinea. (Amber-coloured snail.) Wood, Ind. Test. pl. 35. fig. 172.

In rivulets about Downton, adhering to aquatic plants.

H. tentaculata. Linn. à Gm. Wood, Ind. Test. pl. 35. fig. 176.

This species occurs in abundance in rivulets immediately communicating with the Avon, but I have not often seen it elsewhere in this district.

H. Vortex. Linn. à Gm. Wood, Ind. Test. pl. 33. fig. 40.

On aquatic plants, especially bull-rushes, in rivulets near Stratford.

MYTILUS—MUSCLE.

M. anatinus. Linn. à Gm. Wood, Ind. Test. pl. 12. fig. 35.

The only stream in which I have observed this species, in the county of Wilts, is one near the road from Salisbury to London.

NERITA—NERITE.

N. fluviatilis. (River Nerite.) Linn. à Gm. Wood, Ind. Test. pl. 35. fig. 26.

In the Avon, between Salisbury and Stratford.

PATELLA—LIMPET.

P. lacustris. (Pond limpet.) Linn. à Gm. Wood, Ind. Test. pl. 37. fig. 56.

P. oblonga. (Oblong limpet.) Linn. T. vol. viii. p. 233. Wood, Ind. Test. pl. 37, fig. 37.

I have found both the above species in rivulets, among the meadows to the right of the London road from Salisbury. They are usually adhering to aquatic plants.

TELLINA—TELLEN.

T. amnica. (River tellen.) Linn. à Gm. Wood, Ind. Test. pl. 3. fig. 19.

This shell was first described as an English species (in the 3rd vol. of the Linnæan Transactions) by me, who discovered it in chalky parts of the bed of the river Avon, near Salisbury. It so nearly resembles one of its congeners, *Tellina cornea,* that it is liable to be confounded with that species, but is distinguishable by the position of its hinge, which is not in the *middle* of the valve, as in the last mentioned shell. W. G. M.

TURBO.

T. biplicatus. Linn. Tr. vol. viii. p. 179. Wood, Ind. Test. pl. 32. fig. 145.

Found at the roots of trees, in moss, at Alderbury.

T. elegans. Linn. à Gm. Wood, Ind. Test. pl. 32. fig. 118.

On dry banks, by the sides of the road leading from Milford to Clarendon, enveloped in moss.

Printed in the United States
By Bookmasters